ON EXTINCTION

Melanie Challenger

On Extinction

How We Became Estranged from Nature

COUNTERPOINT

BERKELEY

On Extinction
Copyright © 2012 by Melanie Challenger

Library of Congress Cataloging-in-Publication Data is available
ISBN 978-1-61902-018-4

Typeset in Bembo by Avon DataSet Ltd, Bidford on Avon, Warwickshire
Printed in the United States of America

Lines from 'La Figlia Che Piange' in *Prufrock and Other Observations* by
T. S. Eliot, © The Estate of T. S. Eliot, from *Collected Poems and Plays*
(Faber and Faber Ltd, 2004), reprinted by permission of the publisher.

COUNTERPOINT
1919 Fifth Street
Berkeley, CA 94710
www.counterpointpress.com

Distributed by Publishers Group West

1 3 5 7 9 10 8 6 4 2

There was a whispering in my hearth,
A sigh of coal,
Grown wistful of a former earth
It might recall.

I listened for a tale of leaves
And smothered ferns,
Frond-forests, and the low sly lives
Before the fawns.

Wilfred Owen, 'Miners'

Contents

Illustrations

Beginnings

Natural History Museum, London

An albatross dips towards the sea, then lifts again, beating its wings as if repelled by the opposing magnetism of the water. At first, nothing else stirs. The sea is deathly calm, spread out like a cerecloth. Then a giant rocketing breath hurls a rainbow into the air. As the whale arches and begins to descend under water, the dimpled grey of its back turns into the blue sheen which earned it its name. I see the whale more completely in my imagination. In reality, each sight of it is a jigsaw piece. The whale is simply too huge to be viewed in its entirety. It disappears and, moments later, surfaces at a great distance, its blows sounding notes of both discontent and deliverance.

Watching the blue whale that day, I questioned what it was that I hoped to capture in writing this book. I had been travelling for several months through the southern waters where the blue whales live, determined to understand why these and other marvels of nature were imperilled and why that should matter. Reflecting on the dreamlike moment of the creature's fleeting show, I saw myself as a child again, staring up at a model of a blue whale suspended from the ceiling of a museum.

I first visited the Natural History Museum in London when I was about nine or ten. The architect, Alfred Waterhouse, scaled the building to the dream of diversity, its curves decorated with Daedalian sculptures of living and extinct forms of life. It was a panorama of metamorphosis, a treasury of the wonders of how life evolved and the sobering realities of how life could end. In the alcoves and galleries that lined the echoing building like confessionals were the serried remains of extinctions. The guttered brow of a Neanderthal skull, the bevelled flint tools of early human ambition. Beyond these, the brittle silhouette of *Archaeopteryx lithographica*, the feathered and fearsomely toothed 'strange bird' crucial to Darwin's defence of evolution. The palaeontologists of the previous generation were convinced that the entire class of birds sprang suddenly into existence, whereas the discovery of the first archaeopteryx fossil gave evidence of their painstaking transition from carnivorous dinosaurs. The ancient bird lived on the Earth 150 million years ago, its wings tipped by two large claws, its body flaunting a long lizard-like tail. The discovery of this beast, Darwin wrote, proved more forcibly than nearly any other find how little his generation knew about the Earth's former inhabitants.

Near to a display of the armoured anatomies of trilobites was an ichthyosaur skeleton scoured from its hiding place in beds of shale by Mary Anning, the fossil-hunter who gained notoriety in the nineteenth century for several spectacular finds. She kept her family from debt by hawking the remains of extinct life in her glass-fronted shop in the English seaside

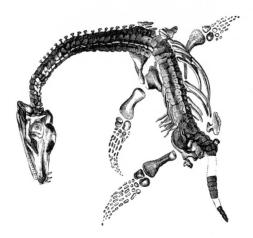

resort of Lyme Regis. She was led to her profession by a local character called Captain Cury, the 'curiosity-man' who stole on to the coaches that stopped at Lyme Regis on passage between London and Exeter. He touted fossils given fashionable names to make them more appealing to female clients: *Ladies' fingers, Crocodile's backs, John Dorie's petrified mushrooms.* After her father fell to his death from nearby cliffs, Mary began rummaging around the seashore in search of her own curiosities. She was only ten years old when she sold her first ammonite to a wealthy lady for half a crown.

There were the grey splintered remnants of a toxodon. On the voyage of the *Beagle*, Darwin and his companions witnessed the unearthing of one of these, 'perhaps one of the strangest animals ever discovered'. A native of South America, the flexed bow of its teeth was like that of a rabbit, while the angles of its eyes, ears and nostrils seemed to ally the giant animal with the manatee. Beyond the toxodon, there was a skeleton of *Mammut americanum*, scraped out of the dusts of Missouri, its tusks trained upwards like bugles to the skies.

Ceiling panels depicted plants both day-to-day and exotic. These gilded illustrations reflected an era when the flowering world was a source of sentimentality and fascination. As specimen-collectors brought seeds and cuttings from all around the world, the wealthy of the eighteenth and nineteenth centuries rushed to embellish their gardens with striking and tropical species, while glasshouses and public parks opened up to exploit their new fancy. Some imported plants became widespread, while other species became rare, their delicate replicas floating above the exhibits as memories of an old season.

I visited the museum again not long before beginning my travels, a nostalgic retracing of youthful fascination. The sturdy wreaths of a diplodocus's vertebrae patterned the floor like footprints mocking its former animation. The fogged glass of the windows filtered the winter sunlight. It shrouded the hall in a cadaverous yellow. A knot of schoolchildren muscled one another for a peek at one of the oddest animals in the hall, the glyptodon, a giant armadillo most probably hunted to extinction by humans thousands of years ago in South America. The beast's small mouse-like head and tail like a huge ice-cream cone softened the ferocity of its bulging hauberk. The children chattered together about the possible revival of such monstrous forms. 'We can bring them to life again,' they said. 'We can make a zoo of monsters!'

In the shadows of the museum, the brooding shapes of earlier creations unsettled the assurances of our tangible human world. It was here that I first encountered the slightness of

human life. Beneath the magnetic appeal of the giant creatures and plants was a terrifying suggestiveness, the millions
of years since their demise. Looking at them with a child's
eyes, I experienced a kind of mental vertigo at the abyssal
distances that lie before and after our brief lives. It was akin to
the rising sickness I sometimes felt while lying awake in the
darkness, contemplating the existence of my parents before
the arrival of me and my sister.

Darwin described a related vertigo on encountering the
giant mammals that once inhabited South America. How
might one reflect on the Americas without a sense of disbe-
lief and awe, he asked, when formerly the continent swarmed
with great monsters? Darwin's theories on how the diversity
of life on Earth arose fortified an old idea that something
larger than our own reality haunted us. The idea was there
in the ancient Greek myth of the *Hekatonkeires*, three colossal beings with a hundred hands each, who ruled the world
at the beginning of time. These giants were the offspring of
earth and oceans, their spirits expressing themselves as forces of nature in earthquake, hail, storm and eruption. Those
who came in their wake eventually conspired to banish them,
fearing as they did their volatile powers.

But the chief touchstone of my awakening to extinction
hung in the Large Mammals Hall, built in 1934 to house
the skeleton of a blue whale that beached on the shores of
Ireland's Wexford Bay at the close of the nineteenth century. Below its bones, strung in the air as an aide-memoire to
mortality, was a full-size model of the whale. Each time that

I visited the museum, I nagged my father to take me back to the exhibit. The blue whale, in large part due to its gigantic proportions, had become an obsession of mine. Entering the hall, I stared across this colossus of saxe-blue flesh and the perspective of bone pinioned to the ceiling, the air transmuted to ocean. Over the years, I never tired of the exhibit. Only when my sensibilities matured to adulthood did I begin to shrink from it. I saw it now as a harbinger of extinction, one of the museum's *reliquiae*, a totem of impermanence. Years later, I would write one of my first poems about the beast's hold on my imagination:

> . . . *she's heavy*
> *In the air, drowning beneath her fathoms,*
> *The mighty eloquence of her breast-stroke, attuned*
> *To the echoes of waves, leadens to extinction.*

Throughout the twentieth century, commercial whaling drastically reduced the population of the blue whale, forcing the beast to the brink of extinction. Gazing at its sleek magnitude, sculptured by the boundlessness of the waters, I did not want my generation to become the last for which this model could evoke the reality of the animal, still living in the ocean, where its exquisite shape evolved over millions of years.

Darwin thought it likely that a raft of natural laws gradually drew the spectacular range of species out of a few, elementary originals. 'From so simple a beginning,' he said, 'endless forms most beautiful and most wonderful have been, and are being, evolved.' But why? To what advantage? According to the Bible, God created the great whales, along with 'every living creature that moveth'. We had no religion in my household and I found the idea that an undetectable consciousness was the source of life's range and singularity profoundly unsatisfying. If this supernatural intelligence chose to create all the forms of existence on Earth, what stayed its hand from saving them from destruction?

From the beginning of time, life proceeded and foundered through no sensible motive. The first geological era of the Earth, the so-called Precambrian time, stretched from the origin of the world to the oldest fossils, the prokaryotes, single-celled organisms without core or genetic history, to the first primitive animals patterned by multiple cells that took shape in the later stages of the era. The shift into Cambrian time signalled the arrival of animals with skeletons, epitomized by the darkly toughened bodies of trilobites that seethed across ocean floors. These and other entities endured mass extinctions unavoidably and for reasons associated with the loss of conditions to which they were accustomed. Extinction was a consequence of huge spans of time, where transformations were mindless processes – an asteroid from outer space, a spilling inferno of molten rock from the netherworlds of the Earth, or the unimaginably slow movement of land masses over one of the magnetic poles – each leading to altered landscapes in which the earlier adaptations of life forms were

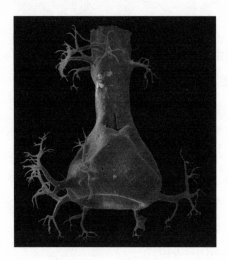

no longer effective. Slowly, the range of a species shrank and its numbers dwindled until they were gone; sometimes a hint of their former glory was preserved in the rocks and sometimes they left an undreamed-of absence. In the great swathes of time given to the Earth, did it really matter if some forms of life died out? Across these spans of almost imperceptible evolution, other entities always emerged in the place of those that perished. What of the improbable bottle-shaped chitinozoa that shrank through the aeons as their outer walls gained spurs, loops and other embellishments, until they disappeared from the fossil record altogether? What creatures might have nosed forth from their odd structures? And of what animal did they, themselves, form some elementary, perhaps almost unrecognizable stage? Did it matter that a great episode of climatic change, in which the whole world seemed to become a glinting prophecy of the Earth's poles, bullied these mysterious life forms into nothingness?

Every great change of climate must be fatal, said the

nineteenth-century geologist Charles Lyell, to those that can neither escape elsewhere nor survive in their transformed reality. For Lyell, it was as if some kind of force majeure acted through the passage of time. During his day, there were several competing theories to explain the physical environments of the Earth. Some believed that observable ongoing processes shaped the world, such as the hourly assault of tidal sea waters on coastlines. Others credited sudden and catastrophic events for the appearance of different landscapes. The undoubting religious believed in the role of a supernatural intelligence. Throughout this period, miners, geologists, enthusiasts and prospectors cracked deeper and deeper into the rocks, increasing understanding of the world's history. Beneath the Earth's swarming countenance were dozens of graveyards, dark undercrofts of generations of life.

On 27 January 1796, the young naturalist George Cuvier laid the tusks of mammoths before an audience at the Institut National de France as proof of extinction. Cuvier used the bones of these elephantine mammals, rescued from the oblivion of Siberian ice, to shatter convictions of the achieved world of a supernatural creator. 'Life on earth has been frequently interrupted by frightful events,' he told them, soberly. 'Innumerable organisms have become the victims of such catastrophes. Invading waters have swallowed up the inhabitants of dry land; the sudden rise of the sea bottom has deposited aquatic animals on land. Their species have vanished for ever.' His audience sat aghast. While Cuvier insisted that the mammoth was a distinct and long-lost species of giant mammals that wandered across the land masses of North America and Eurasia thousands of years earlier, his contemporaries believed their tusks were those of living elephants. Cuvier regarded these blunders as the consequence of deficient and unresolved scientific thinking. The studies of elephant bones published by his contemporaries, he explained, were so insubstantial that one could not decide whether they belonged to a living species or not. For Cuvier, the disappearance of the mammoth argued for catastrophic natural events in the past that irrevocably altered life on the planet. People were appalled, and so they doubted. As Thomas Jefferson exclaimed in 1799, puffed up with such belief in the bounteousness of America that he could not countenance the idea of these losses in its history, 'If this animal once existed, it is probable on this general view of the movements of nature that he still exists.' But the ancient rocks gave up dark imprints of bone after bone that eventually made extinctions, if not the catastrophes that might cause them, an inescapable conclusion. Layers of rock had captured the ghostly outlines of many species that no longer walked the Earth, including what appeared to be primitive forms of man and the remains of his early arsenal.

In a letter written to John Stephens Henslow while on the *Beagle* voyage, Charles Darwin described his discovery of a mastodon, an enormous and extinct relative of the mammoth. He also happened on the teeth and lower jaw of a huge ground sloth alongside the remains of numerous giant creatures. Darwin did not directly propose the human hunter as the culprit in the disappearance of these lumbering creatures, but others writing in the decades before and after the publication of his theories were more explicit in their claims. Contemporaries like Scottish zoologist John Fleming recognized that the visible effects of humanity on the world's species demonstrated our potential to be the primary cause of extinctions. Whether a savage or a nobleman, Fleming claimed, man was powerfully motivated to pursue a destructive campaign against his fellow residents on the globe. In 1825, Professor Joel Allen published *The History of the American Bison*, which mourned man's ruthless extermination of the buffalo throughout vast stretches of the continent. Contemporaries like Miller Christy, whose census of the species found only a thousand remained, backed up his work, especially as it emerged that in the years following the construction of the transcontinental railways, hunters slaughtered huge numbers to meet the increasing opportunities to trade skins. In 1848, the dodo bird of Mauritius, the first icon of extinction by humans, became widely known through the publication of *The Dodo and its Kindred* by the English naturalist Hugh Strickland and his colleague Alexander Melville. And then, in the latter half of the nineteenth century, George Perkins Marsh published his account of the effects of man on the natural world, which gave further grounds for a new way of considering extinction. 'Man is everywhere a disturbing agent . . .' he wrote. 'Wherever he plants his foot, the harmonies of nature are turned to discords.'

★

A little under twenty years after my first visit to the Natural History Museum, I stood crouched inside a gigantic rusting cast-iron vat, one corroding feature of the global whaling industry that endangered the blue whale – an industry now, itself, well-nigh extinct. Grytviken, the first shore whaling station in the Southern Ocean, lay in ruins before the tattered mountains and mewling glaciers of the subantarctic islands of South Georgia. The colours of the landscape were blunt, decisive. Reds, whites and rich, mineral greys, and a gash of blue as if a strip of cloud had been torn away to reveal the bright sky beneath. Memories of Grytviken's violent past eroded across the glacial scene, steadily breaking down into nonsensical fragments. I felt overwhelmed by the desire to make sense of the decay, to salvage the relevance of the

place before the censoring actions of snow and ice buried it.

My old, half-forgotten fascination with the blue whale had led me here, to the slaughtering-grounds of thousands of whales. But now a different poignancy tainted my interest. Many of the life forms on display in the Natural History Museum died out in mass extinctions during ages in which our species didn't even lurk in promise. Huge extinctions like these were terrifying, exceptional events in the ancient past. But in the years after those early visits to the museum, I became aware that I was living through another mass extinction of animals and plants without even knowing it, this one due to human behaviour. I wanted to explore the idea of extinction in the light of this new, sobering reality.

Before the journey that took me to Grytviken, I was living in Cornwall, near Penzance, a town at the south-west extremity of Britain, a stormy place ringing constantly with the sound of boat masts in the wind. Rows of granite buildings and Georgian merchants' houses peered out on the Atlantic Ocean, lines of profit strung from the sea. The whole landscape expressed the once intimate but now almost entirely broken relationship of the inhabitants to the natural world. Everywhere across the county were relics of this former closeness, visible in crumbling structures and discarded tools, in crooked patterns imposed on the landscape. I began to think about what these ruins signified, the losses inherent in the deft sufficiency of modern life. It was here that I was confronted by my own fragmented connection to nature. I began compiling notes on the subject of extinction, from the philosophers who grappled with the purpose of life and concluded that humans lay at its heart, to the poets of the eighteenth and nineteenth centuries who, faced with the lightning appearance of industrialization, saw nature as the antidote to the corrupt forces of civilization. From Aristotle to Ralph Waldo Emerson, Cicero to Percy Bysshe Shelley, the way people viewed their relationship to nature affected how they lived with it – destructively, admiringly, thoughtlessly. I began to question whether my own ignorance of nature was associated with the damage societies wrought on it.

My chief interest was in gathering a history of how we had become so destructive to the natural world and its diversity. But this curiosity, in its turn, sowed the seeds of other questions. What were our emotional responses to disappearances? And what did such responses mean for the kind of animal we had been in the past or might become in the future? My search for answers took me from Cornwall to the freezing regions of Antarctica and the Arctic, and from a small, extraordinary, uninhabited island in the archipelago of the Falkland Islands to the lively assemblies of New York. Although separated by thousands of miles, the unique histories of these places kept company through the ages, each touching on and deepening the matter of extinction.

The First Peregrination

West Penwith, Cornwall

1

Wild flowers

Although archaeologists uncovered evidence of early man's potential to exterminate vulnerable wildlife, the endangerment or extinction of species began to accelerate in the later stages of human history, around the time of the voyages of Captain Cook in the eighteenth century. As industrialization progressed, Europe lost a staggering proportion of its natural habitat, which restricted and in some cases eradicated many species that relied on such wetlands and grasslands, forests and waterways. The pattern of destruction spread elsewhere. During my childhood, I came across a book in my father's study, a large burgundy hardback, its leather spine gilded with the title, *Longman Illustrated Animal Encyclopedia*. Its pages infused me with the fever of creation. I spent years absorbed by it, tantalized by the extraordinary, dangerous and beautiful kinds of life that existed at the same time as me. The *Linophryne arborifera*, a small blue fish dwelling in the depths of oceans with a seaweedy beard and a ghostly false eye hovering above its head. The legless skink of Madagascar, with visor eyes shielding it from dust as it burrowed. The two-legged worm lizard of Mexico, with its long, corpulent tail. Beside some of the descriptions were the letters E, V and R, which stood for Endangered, Vulnerable or Rare. The book's introduction explained that these symbols derived from the findings of the International Union for the Conservation of Nature (IUCN), founded in 1948. 'Many of the world's animal species are in danger of extinction,' the author stated.

Those bizarre and magnificent animals threatened with extinction began to intrigue me the most. The pink-fairy armadillo of Argentina, a tiny, stubby creature, mantled in shell-pink armour. The peculiar aye-aye, one of a kind, its long, searching fingers adapted for life in trees. The axolotl, with scarlet eyes and flimsy pale pink legs and feet, feathery gills and a blue blade of tail. As I looked at the illustrations, I felt a simple indignation that such odd and unrepeatable beauty would cease to exist.

During the latter half of the twentieth century, the IUCN began gathering information about the world's species of plants and animals. The researchers discovered that over-fishing, pollution and coastal development were forcing numerous sea creatures into a critically reduced state. The destruction or alteration of habitat through agricultural intensification, construction for tourism and industrial expansion was endangering a third of amphibians, more than one in eight birds and nearly a quarter of all mammals investigated. Water extraction and pollution threatened freshwater fish around the world, especially in places like southern Africa. The deliberate or accidental introduction by humans of alien species overpowered huge numbers of native species. Industrial logging in tropical regions put species of dragonfly and damselfly at the risk of extinction. In the oceans, more than a quarter of the world's reef-building corals were on the brink of disappearing. These gnarled underworlds of many species of invertebrates and fish were foundering due to bleaching and disease as coastal development and pollution escalated. The world's forests, especially stands of conifers, and cycads, those living, flowerless fossils that millions of years left untouched, endured significant threats from agriculture and logging.

While I began to despair at our species' ruinous potential, I also experienced hope as our expanding moral concerns included a willingness to protect the life forms living alongside us. Some animals such as *Equus ferus*, the wild horse, had come back from the brink following the efforts of conservationists. The fates of those creatures that had mesmerized me as a child were now monitored attentively. The pink-fairy armadillo was protected under regional and national legislation in Argentina, guarded inside the Lihué Calel National Park. The habitat of the axolotl was being rejuvenated through the Parco Ecologico Xochimilco. I was puzzled by what prompted such acts of salvation in an otherwise lethal species.

Extinctions appeared to fascinate us, and I was interested in exploring this fascination and its relationship to our capacity to limit our destructive potential. For me, these were clues to our character as a species and our affinity with nature. I began to consider the haunting or arresting signs of mutability in the landscape and wondered why they aroused people's curiosity and filled them with tender feelings. I became convinced that the nostalgic sensation many people experience for things both disappearing and eternally lost might prove essential to fostering a more favourable approach to nature.

I was then living about four miles from Penzance, working in a cabin on Ding Dong Moor, a small stride of tumbledown farms, mining cottages and Neolithic ruins. The cabin was at the bottom of an unbroken strip of ground, the once clipped order of an old garden still detectable through its wildness. By the roadside stood the striking remains of a tower-like building, its glassless windows framing the distant and fitful sea. Feuding elements of wind and rain had reduced the stone to rubble and slime. A signpost in the lay-by identified the structure as the ruins of Ding Dong mine, one of the oldest tin mines in Cornwall. Years had obscured the origins of the mine's name. Through the long winter months, gales unleashed the sea waters from their confinement, demolishing any sense of ocean and sky. Mists steeped each day, hushing up the Earth. This old building and its history were becoming naturalized to the soils. In time, there would be nothing left but the eye's uncluttered view of the moors and a nonsensical name on a map.

> *Ding-dong! Merry, merry, go the bells, ding-dong!*
> *Ding-dong! Over the heath, over the moor . . .*

The wrecked mine came to focus my first ideas about extinction because it represented both the forces involved in our shattering effect on nature and the regret inspired by loss. Inside the cabin, there was a makeshift kitchen, a faded floral sofa bed and a desk. Rainwater had soaked through the window frames and crept down the panes, spreading into little deltas on the wooden sill. On the far side of the desk, I balanced a mound of books to read during my stay, including two claret-coloured leather volumes of English poetry. My view from the desk was of the defeated garden, the moorland snarled by bracken and gripping mists, the faint unrest of the sea. For the first few weeks, the weather was relentlessly grim. The rain, dense and sooty, struck down the view, sealing me inside the cabin and my thoughts. During this confinement, my eyes would leave my page and stare absent-mindedly at the landscape before me. When thoughts resurfaced, they were usually concerned with what I'd been reading moments before. The ancient granites that propped up the moorland, the tough thrust of Cornwall in which the minerals of the region's former prosperity had lain. Three hundred million years ago, these rocks boiled up to domicile in this landscape. *Three hundred million years ago!* A world in which early forms of insects still dreamed of flight and the boomerang skull of the newt-like *Diplocaulus salamandroides* was not yet a ghost in the grit.

Time bound meshes of natural history into the thickness of these ancient rocks. Curiosity or enterprise led people to rupture them, unsealing ancient realities that revolutionized their minds. These rocks had pushed up through the border of the Permian and Triassic eras, when the unstoppable progress of mass extinction eradicated nearly all of the era's species, disappearances that were then succeeded by the erratic inventiveness of living cells. Their ceaseless innovation led to new forms – those strange anatomical conclusions of nature that captivate children's imaginations. For example, the small amphibian *Diplocaulus salamandroides*, whose pronged skull seemed to want to split into two minds, faintly echoed in the hammerhead shark, a beast that fascinated me so thoroughly as a child that I implored my mother to draw it over and over again, as if each trace of its silhouette might vindicate its freakish proportions.

After a few weeks of confinement, jitteriness got the better of me. I abandoned my books and stepped outside, ignoring the blustery weather. Bundled in waterproofs, I left the cabin and took the path across the heath that bordered the ruined engine house. It was an early morning in March. Mists swung between the hedgerows, homesick for the seas. The wintering birds sang through a dawn still struggling free of the cold

months. The tiny shadow of a solitary buff-brown bumblebee fled across a bed of nettles. The moor wheezed and snapped with haphazard actions of survival and a recent fall of rain weighing on the tangle of plants and undergrowth.

I was acutely aware of being a stranger to the moorland. I had no words for the sounds that I heard. In my ignorance, each birdsong entered my consciousness as a sweet but secret music. All clues as to the type of bird were beaten into the background. Each bee was just a bee, small and sombre, directionless. I wished I could muster the words for the things I saw and heard in an hour or so's brief ramble – the creatures that showed themselves and the telltale signs of others shyer of my footfall. The plants that flowered and thorned, the tiny green promises of the coming season. But I couldn't, without falsifying the memory. I was bereft of speech for this landscape, suffering from a kind of amnesia shared with others of my generation. There was so much we didn't know about the natural world around us. What were the tiny birds that pinged out of the bracken as if my step triggered them against their will? Did these little chirring creatures live inside the moor's prickly clutch all year long? Or was their indignant flight a harbinger of spring? I could see acres of nameless, incoherent greenery. When would the pluckiest wild flowers appear in defiance of late frosts? And what seasonal eruptions of colour would alter the complexion of the moors throughout the year? My perception confined me to the present.

I recognized gorse and bracken but in the careless, almost indifferent way that I could put a name to daisies or buttercups.

I didn't know their influence on the other species or the plants most likely to flower beside them. Why should I? Somewhere in my childhood, I'd learned from the constitution of my environment that the natural world and human nature somehow warranted separation and that each existed, perhaps even existed to advantage, without the other. But the more it struck me just how little I knew about this landscape, the less my ignorance seemed obligatory or reasonable.

Extinctions were visible across the moor, the shabby outlines of ways of life now long forgotten. If I'd had insight into the landscape, I might have been sensitive to other extinctions that had left no trace – quiet once filled by particular birdsong, the deadened hum of an insect, the stripped earth of a familiar flower. All I could grasp were signs of human activities that were now obsolete. Old pathways blurred by mud through which farmers once chivvied their grazing animals, scruffy hedges fouled with rusting wire. There was a stone feeding trough, purloined now by birds, their brown heads bowing to the water like worshippers. A gauze of bushes and moss stretched over ancient field systems, greying into the distance. There, the dark back of the sea slowly heaved. The soft scars of previous generations' farming hinted at losses far greater than I could understand, memories and skills that had vanished, the discrimination of seasons and assessments of the land's potential that crushed or occasionally assisted the abundance of nature. I paused outside an old ruined farmhouse, the rain knocking futilely on its entrance. Ivy had spread across the face of the house and on to the roof, a slow green wave of interment. Staring at the cracked windows battered into blank holes by the wind, I wondered who had once worked the land here. And how the place would have appeared when still a working farm. Why had the farmers abandoned it? Did anything remain of their knowledge in the lives and experience of their descendants?

Even within my own family, there once existed a greater recognition of the natural landscape. Somehow, it was not passed on to my generation. My grandmother seemed to have an inborn sense of what had been forsaken in the decades between us. For as long as I knew her, she experienced a powerful nostalgia for the rural surroundings of her childhood, the source of her greater sensitivity to nature. She recorded her feelings in a large, pale green notebook, the title, *Book of Memories*, softly smudged in blue and red crayon. On the first page, she listed the births of her family members, siblings whom I had heard about but never met: Dorothy, Marjorie, George. By each of these was the name Kingsclere, the village in Hampshire where they were all born. She spoke of it as an unspoiled, peaceful place, a rustic landscape whose faults and misfortunes had been softened by time.

John Porter, the most successful horse-trainer of the Victorian age, wrote an autobiography in which he described the flowers of Kingsclere: harebells, wild hyacinth, wild thyme and saxifrage. He regarded it to be a 'wonderful county for birds', noting the presence of 'warblers and whistlers and twitterers', and particularly the grey plover, the gentle augur of spring. Porter had his stables at Cannon Heath, near Kingsclere, where there was a rookery, and he often saw kestrels and heard the cry of curlews, the 'jug-jug' of nightingales. The 'rustic merriment' of an agricultural fair took place twice annually, and there was a market of fresh, local produce every Tuesday. Porter wrote the book in the early nineteenth century, when Kingsclere was an enclave of cottages with wooden dormer windows and gardens bounded by flint walls. The ancient Norman custom of the curfew bell still tolled from the old church. There were wheelwrights' shops and saddlers, and general stores selling everything from flour and bacon fat to starch and powder blue. In those days, a Kingsclere breakfast was a hearty confection of trout, steak and strawberries.

My grandmother's family moved to the village just before the First World War. Horses were still the dominant means of transport and she noted in her *Book of Memories* that 'Aunt Ada could drive a four-in-hand easily.' I tried to imagine the gruff breaths of the horses, the seasoned stench of their ordure. Her father owned a grocery and bakery on Swan Street, one of the chief roads through the village. Fresh produce was supplied by nearby farms, which, along with several watermills, blacksmiths and a tannery on North Street, employed many of the villagers. The farms provided milk and cheese, along with meat and skins from the herds, and timber, which was sawn locally. Butter was churned on the farms and chilled against the dank edges of wells. Specialized workers met the needs of the farmers, such as a Mr Bennet, who fashioned by hand halters for horses, reins for ploughs, and pig nets. I could remember my grandmother's descriptions of the harvests, when rows of brawny horses pulled binders, hugging together the tow-coloured corn. She mourned the loss of the farming land, much of which disappeared under roads and houses in the years after her infancy, although she did recognize that it was a hard life for labourers. Workers often rose long before dawn and walked several miles, the demand for sleep hanging like ballast from their bodies, slowing their gait, before starting the working day on the farms. But despite the hardships, she believed her generation possessed quite effortlessly some things that had disappeared from our lives. My grandmother lived half her childhood outside, where she felt safe and confederate somehow. She understood that everything she and her family possessed relied on the dormant fertility of the Earth, the scramble of light, rainfall, birds, insects, worms, flowers.

What she spoke of most were the days spent in Hawkhurst woods, close to the village, and of the wild flowers that grew there. Her daily excursions through the fields endowed her with her knowledge of natural history. Doubtless, she was less conversant with the natural world than the generation before her, but she passed on what she knew to my mother and her siblings. Her favourite wild flower was the common dog-violet, whose delicate blue flash appeared in the fields that led to Hawkhurst woods in the early summer and again at the summer's end. Doubtless, my mother knew less about the flowers and grasses than her mother did and spent fewer hours outside. Yet she still felt certain about her favourite wild flower, the bluebell. I knew little of the wild flowers that grew near my home, whether rare or common, and nothing of their season's rituals. My mother sought to pass on her sensitivity to our native flora to my sister and me on our walks in the woods and across the Berkshire Downs. But she later spoke of a strange struggle, of outward causes and influences that overpowered her own ordinary, motherly energies, a lessening, perhaps, of the public prominence given to this knowledge, which somehow sealed off our minds. And so, as I looked out across the moors and asked the question of myself, I realized I didn't have a favourite wild flower. I simply did not know enough of them to make the choice.

In my grandmother's old age, this bucolic early history was so vibrant for her it was as if the trees and flowers grew again, taking the place of her shadow. Her mind was almost gone, crumbling through dementia each day into the most stubborn or cherished parts of herself. But I could still picture her, animated and intact, in the dark kitchen of her terraced house, dressed in a blouse, a pleated skirt and an apron, with a dead pheasant on her lap that she was plucking for dinner. Her fingers worked deftly and swiftly, a soft haze of down besieging the movement. She must have gone to a local butcher to fetch the bird. I had only ever known the skinned meat that we bought from the supermarket. And so I sat on the chair beside the battered, old-fashioned cooker and watched my grandmother intently as she worked. Occasionally, she took hold of the pheasant's limp, cleric's neck and raised its scarlet head to tease me. Mostly, she sang old tunes or chanted common nursery rhymes in her sibilant voice:

Little Boy Blue, come blow your horn,
The sheep's in the meadow, the cow's in the corn.
Where is that boy who looks after the sheep?
Under the haystack fast asleep . . .

She never gave any sign that she recognized the border that separated our experiences.

I grew up in a town a few miles from Kingsclere, a backwater blighted by its closeness to London. Trade routes cordoned its regions, from the river, which attracted the first farmers thousands of years ago, and the canal, constructed in the eighteenth century for narrowboats transporting coal and goods, to the railway, which opened in 1841, bringing an end to the brief heyday of the canal system. By the time my family moved to the town, when I was a few years old, several motorways and major roads strangled it further. As a consequence, I lived with the feeling that commerce immured me, shutting me away from nature. I felt suffocated by the characterless shopping centres, the modern housing estates, and the constant tattoo of traffic left me with a slight sense of dread. Whereas I could not wait to escape, my grandmother's memories fixed her there, longing for the way it had once been. Her memories overlay the landscape, haunting it with things I could not see or understand.

Many of the changes that occurred through her lifetime subdued or annihilated the natural world, replacing it with man-made materials, noise and technology. Perhaps my grandmother's nostalgia was in part an angry rejection of the pervasiveness of human society and a sense of impotence against powers of transformation far greater than her own. I was struck by the wistfulness of her portrayals of her past in Hampshire, a rural landscape safe from the inroads of traffic and enterprise. Unable to encounter the landscape as it once was, I had to rely on her insistence that a more luminous, expressive countryside once existed. But could I trust her vision of the past? Should I steer myself towards it? Or, once there, might I find a world in its own way as stifling and threatening as my own? It seemed to me that my grandmother's nostalgia emanated from a compulsion, a deep-seated need. She idealized her childhood in Kingsclere; she polished the dawn birdsong, she subdued the wind. Everything was softened, sifted of its jagged or painful elements. Yet the comfort of her old age was made possible by the giant strides that had shattered her childhood world. She rued the decimation of wildlife, but she also longed for the earlier methods of farming that were an essential agent in this progress. The adversities that led to the changes that progressively destroyed what she held dear were hidden, like a cancer that corrupts the body.

A week or two later, I passed the owners of the cabin, Doug and Jenny, who invited me into their home for a cup of tea. The room was dark but heartening with a large wood-burning stove beside which an ancient-looking cat lay curled, its grey fur matted into motionless peaks like a frozen sea. Jenny's long, straight hair reached to her waist. Her hooded, soft blue eyes gave her face an expression of gentle sleepiness. As we chatted, I tried to explain the work I was doing. 'I'm trying to think about extinction,' I said. A shimmer of affinity passed across Jenny's face.

'There's a long history here,' Jenny said, 'and it's threatened. We're trying to regain control as commoners but we don't know how long we can save this landscape.'

Over many centuries old agricultural cycles of grazing and burning encouraged the wildlife and plants now long established on the moors, she told me, from mauve-freckled orchids that popped up in the revitalized soil, to the rare high brown fritillary butterfly. Reptiles like the adder and slow worm, whose numbers had dropped over the past decades, relied on the stunted clinch of the heaths and moors. It was a man-made landscape, yet it harboured a wild blend of creatures and plants. The endurance of the moor, she argued, was contingent on the concern and active participation of those living in the area.

'It's quite complicated,' she said. 'Most of the commoners in the area are sympathetic to grazing and together we've got some chance of managing things for ourselves. But it's always uncertain, especially if someone wants to sell their land.'

It was a controversy that stretched back over twenty years. In the 1980s, some people living close to the moors became concerned about their threatened disappearance. On 25 October 1984, a meeting in St John's Hall, Penzance, was entitled The Penwith Moors: A Vanishing Landscape. But the history of man's influence on this landscape can be traced as far back as the Bronze Age, when settlers began fostering the lowland heath of the moors, chopping down primeval forests and using the land for grazing. For countless centuries, men and women kept the heaths alive through paced spells of grazing, burning and cutting. Many of the heaths were absorbed into communal husbandry. Then the whole community was involved in farming and had a stake in the landscape. A feature of such settlements was the common, or 'waste'. The commoners shared these areas, putting their animals out to pasture and taking their measure each season in wood, stone, or sods for fuel.

The poorest in the county, as elsewhere in Britain, were enterprising. They nailed notices to trees and gates, recommending themselves as hairdressers or tailors, selling everything from maggots to homemade sausages. The wealthier classes in England sneered at and ridiculed commoners. Accounts such as *Campania Foelix*, Timothy Noarse's discourse on improvements to husbandry written in 1706, described them as troublemakers, as wild and barbarian, mischievous, lazy, reluctant to work a full week, and liable to sabotage harvests and livestock. Noarse denounced them as 'trashy Weeds or Nettles, growing usually upon Dunghills, which if touch'd will gently sting, but being squeez'd hard will never hurt us'. But commoners persevered, reliant on their intimacy with nature. Across England they kept bees, grew hemp and flax for clothing, trapped rabbits and birds, fished in nearby streams, rooted around for edible flowers, mushrooms, nuts and berries, foraged for wood and watercress, gathered teazles, and wove mats and baskets from rushes – seasonal activities that reflected the natural thriftiness of their lives.

Communal husbandry began to alter in concert with a string of agricultural developments. By the seventeenth century, water meadows improved the quality and permanence of grass, the nitrogen fixation from new crops enhanced soil fertility, and leading animals out to dung restored nutrients to arable fields, acts that steadily boosted harvests. For a long time, too, there was a reciprocal relationship that lessened farmers' vulnerability to fluctuations in crop prices. Landlords wrote off arrears of rent during harder times, passing on the government's land tax to their tenants in better days. Those working the land could then more confidently experiment with efficiency, while landowners assisted such trials by purchasing grass seed and funding the reorganization of the farms. New contrivances and methods were sought to overcome troughs in the market and as the corollary to phases of relaxation and prosperity, when cheapening loaves released spending power. Mounting arrears on land rents always forced farmers to invent new ways of increasing yield and lowering costs, striving to keep the wolf from the door.

The frugal life of communal husbandry became unworkable as the population grew; increasingly it was necessary to reap more resources from individual farms growing crops suited to local conditions. The gentry sought private ownership of the land from the British government, prompting legislation that swiftly extinguished all common rights. The majority of Enclosure Acts were passed by British governments between 1750 and 1860, although piecemeal enclosure took place from the thirteenth century onwards. This partitioning and privatization of land brought the extinction of common-field farming, sometimes destroying entire villages, ploughing through old traditions of sustaining a family from the land. As the Cambridgeshire poet John Clare wrote in 'The Mores',

> *And birds and trees and flowers without a name . . .*
> *All sighed when lawless laws enclosure came.*

Farming became increasingly individual rather than co-operative, the old peasant labourers turned into productive agricultural workers and tenant farmers with conventional wages and less time on their hands.

The commons of my grandmother's home were enclosed in 1854. Born some sixty years after this change, she became heir to a landscape and a people's connection to it that had already altered irreversibly.

Heaths across Britain and continental Europe decayed through lack of interest, undergoing a kind of wasting disease as their appeal or the need for them waned. By the time of the British parliament's Commons Sitting on food supplies in March 1941, many people no longer exercised their rights on common land. The large acreage of commons once grazed, observed Earl Winterton in a conversation with the minister of agriculture, had become overgrown and useless. The minister responded tersely that the committees in charge of agriculture during wartime already had the power to take possession of common land for food production. After the war, many abandoned stretches of heathland began to metamorphose, slowly eaten alive by the dominant species of vegetation. The conversion of some of this overgrown land to more intensive farming was actively endorsed. In reclaiming heaths, farmers destroyed the old and distinctive ramshackle stone-hedged field systems that previous occupants created over millennia, along with the remains of Bronze Age settlements, disturbing vulnerable wildlife.

I looked through the windows at the muggy light of the afternoon. Mists moved against the still hedges, slow exhalations of the failing day. Jenny spoke with a certainty, shared by my grandmother, about the vibrant potential of the landscape, though her words were imbued with the contemporary sense that nature is now inarguably damaged.

But not everyone agreed. The September after I left the cabin, a hundred or so people gathered for a protest march to the Nine Maidens, a prehistoric stone circle near Ding Dong. Their complaint was the erection of fencing planned by the local commoners, Jenny among them, to enable the return of grazing to the area. The group was called Save Penwith Moors and its members were resisting the revival of heath management because they saw this as another artificial means of controlling the landscape. According to them, the landscape was transforming on its own terms and the people visiting the heath gained much more from its rough, unbroken simplicity. It was precisely the dour, undetermined nature of the place, unfenced and unmanaged by humans, and the

decaying remains of earlier and extinct societies, which provided some unexplained but tangible benefit.

Jenny's partner, Doug, was more sceptical too. Having lived
in the cottage at Ding Dong for decades, he felt that the
natural world flourished in unexpected ways: 'As one thing
ends, another enters and thrives.' Doug didn't think that the
landscape was in peril, only altering into something else. People had controlled and changed the natural world outside
Doug's window for thousands of years. Once mining was
profitably exhausted on the moor and the idea of farming
the land was less appealing, the landscape near Ding Dong
was untouched but for the winding paths left by walkers and
tourists. And the movements of those, like Doug and Jenny,
still living in the area. For some, this made for a wild and expressive landscape, poignant precisely because it now seemed
beyond cultivation. Without people cutting gorse for fuel or
grazing animals trampling and eating the undergrowth, the
shrub would burgeon along with the husky, stubborn fern of
the bracken. While this barbed tangle might give shelter to
some species, unchecked it would alter into a scrub landscape
where saplings would sprout, pale green hints of the forests
that once overshadowed the scene.

When winter turned to spring and the mild sun slowly lifted
like a heavy drum, I warmed my back against the weathering rock below the broken window of Ding Dong mine
and read. Between paragraphs, thoughts and thoughtless survey, the life before me began to change. At first, it was an

ambient agitation, a drowsy sense of increased din and liveliness. Gradually, this increase became insistent enough to provoke investigation. *Yes*, there were more birds singing and frisking between the twigs than in previous weeks, the rabbits rumped around the bushes with more gusto, and a single butterfly stood applauding softly on a bramble leaf. Among the coiled greenery that made up the moorland, water resounded anarchically, as an orchestra tuning up through the disordered onset of harmony. I listened, really listened. All that natural life and so good to be among it!

I recalled a line from Charles Darwin's writings on the struggle for existence. 'Lighten any check,' he wrote, 'mitigate the destruction ever so little, and the number of species will almost instantaneously increase to any amount.' The mine was finished. The road was out of the way of ordinary traffic. It seemed that another order of life was alert to the absence.

Restless and excited, I left the mine and continued down the path. It crossed the gorse and bracken and gave way to a rough track that descended towards a quiet, narrow road stitching the area to the outside world. An old two-up, two-down miner's cottage stood fenced and unoccupied. I paused outside the cottage, troubled by its emptiness, fantasizing about who might have occupied the remote abode during the bleak winter months of the past. The distinctive white-painted granite was stained by thick palls of smoke from the mine's winding engine. Despite the absence of human life, I couldn't feel alone; I was immersed in the ceaseless, unguessable gestures of life. Something struck me, beyond the reach of conscious thought but none the less sensible to the body in little rushes of feeling.

The death of the old customs of the moors left a sense of feral jubilation across the landscape. Through my conversations with Jenny, I knew that it also restricted the range of life that once prospered here, pointing towards the eventual extinction of the landscape in this particular guise and some of its wild inhabitants. But in returning the heaths to active agricultural use to safeguard some species, no covenant of permanence was signed, no assurance that those lending a hand to wildlife would always prevail over other possible uses that might suppress or destroy nature. Once a working attachment to any landscape is established, there is always the potential for land use to alter to satisfy people's changing demands.

In the past, those with a close working relationship to a landscape were often unsentimental in pursuing their goals. Throughout the history of communal husbandry, the demands of agriculture excused the extensive suppression of any life form that interfered with its objectives. Across Britain, legal measures and an array of brutal methods like slings, pitfalls, foothold traps, snares and poison led to the destruction of many plants and beasts. In the back of my grandmother's *Book of Memories*, I found a series of letters to the editor of the newsletter for Kingsclere. 'Memories of Long Ago', my grandmother had entitled these. They included a scientific report from the 1930s. Sparrows, the report claimed, consumed around an ounce of corn a day, representing a loss

of tonnes of corn for farmers. Overnight, a bird that farming people already perceived as a nuisance was formalized
as a pest. Farmers drummed up support for 'sparrow clubs',
paying around a penny for every twelve sparrows destroyed.
One club in Kingsclere used a pocketed net on two twelve-
foot canes, with one large cane for a 'rucketer'. I paused as I
came across this word, which was entirely unfamiliar to me.
I failed to find a definition of it and concluded that it must
have been an old word, borrowed or adapted to the practice
of killing sparrows. The sparrow hunters would creep out at
night, drape the net on the canes in front of an ivy-covered
wall, shine a box-shaped hand-held lantern called a bullseye
on the net, and ruffle the ivy with the rucketer. The sparrows would fly out, go for the light and get caught in the net.
The rucketer swiftly dispatched them. The massacre was of
such proportions that one woman in the village, Mrs Priest,
became renowned for her delicious sparrow-pie suppers.
The suppression of sparrows persisted for a number of years
across England, until a new, unforeseen antagonist appeared:
the shambling bodies of millions of caterpillars. 'They were
so dense crossing roads,' one villager wrote, 'that horses and
cars were slipping and skidding on their bodies, and day and
night one could stand and listen to the ominous rustling of
crawling and eating caterpillars.' The caterpillars were those
of the common cabbage-white butterfly, whose numbers
soared with the loss of their natural predator, the sparrow. In
a turn-about, the 1954 Protection of Birds Act safeguarded
the sparrow against direct persecution by farmers.

It made me smile to remember the letters my grandmother exchanged on the matter with someone from the village who chose to remain anonymous, signing himself as *Rara Avis*, rare bird. 'It would be rather nice if you signed your own name, don't you think?' my grandmother had written, playfully.

The fluctuations in sparrow numbers, especially house sparrows, were recorded only when people paid attention to them. But fewer had been surviving to breed in the decades since my grandmother was born. Some blamed this on the shift from horses to motor vehicles, as sparrows used to feed on the cereal given to horses. Others pointed the finger at the mechanization of the grain harvest and the loss of places to nest in towns and cities. For some, the cause was the sparrowhawk, their natural foe, a creature that itself faced extinction in Britain when the use of organochlorine pesticides such as DDT in the 1950s destroyed their eggs. Once such pesticides were banned, sparrowhawks swept back into the skies of towns and cities, preying on their namesake.

The plight of the sparrow suggested to me that nature, like the evasive accuracy of human memory, was sometimes too tangled and dynamic for any simple turning back of the clock.

In the quiet of the cabin, the light began to retreat as if drawn by an imperceptible tide back towards the distant sea. The moon was halved by shadow, its brilliance leavened the darkness with thousands of silvery reflections. A book left by previous occupants of the cabin sharpened my perspective on the loss of knowledge since my grandmother's childhood. The introduction to Frederick Davey's *Flora of Cornwall* was a kind of old-fashioned obituary, written by the author's friend, Chambre Corker Vigur, a physician from Newlyn. According to Vigur, Davey was a handsome, dark-haired Cornishman, who turned to the study of nature after rheumatic fever shattered his health at an early age. Whether as convalescence or compensation, Davey rambled through the fields and lanes of the county in his spare time, disappearing behind mounds of books on natural history on lamp-lit evenings. Davey brought together his observations from these excursions in his treatise on the plants of the region, published in 1909. Vigur remembered him and the days they spent botanizing around Cornwall. Davey, who spent his entire life in the same small parish in West Penwith, discovered a new species of eyebright while out on his walks, which he named after his companion.

Vigur's eyebright, native to Cornwall, its lilac petals pouting like a thwarted child, thrived on the gorse and stubbed grasses of lowland heath. Over hundreds of years, the plant adapted to a landscape disturbed by other species, flourishing in the spaces where cattle crushed the bracken and gorse. But when I searched for it in a general book on wild flowers, I discovered that the species was in crisis. In the years after Davey's death, as use of the Cornish heaths declined, Vigur's eyebright began to disappear. It was already gone from more than half of those places where it bloomed in Davey's day. The neglect of the moors was a chief cause of the likely extinction of Vigur's eyebright. Without inhabitants who needed to use the heathland, a flower of such delicate circumstances could easily disappear unnoticed.

Davey's interest in the plant sprang from a perspective of pleasure, not necessity. There were historical precedents for botanizing of this kind. In 1760, while rambling in Cambridgeshire in April, the poet Thomas Grey noted in his diary the spreading leaves of horse-chestnut, sallow, jonquils, hyacinths, anemones, wallflowers and auriculas. In 1802, Dorothy Wordsworth, with her brother William by her side on a morning walk through the countryside of Westmorland, marvelled at 'Primroses by the roadside, pilewort that shone like stars of gold in the sun, violets, strawberries, retired and half-buried among the grass. There we sat,' she wrote, 'and looked down upon the green vale.'

On the one hand, this appreciation stemmed from the earlier days of local husbandry and widespread rural skills. In the pre-industrial world, there were fewer alternatives to agriculture, even when times were lean, but labourers steadily switched to trades like factory work and framework knitting, lured by the idea of wages not subject to the unpredictability of nature. Flowers and greenery were an integral part of the year's rites and celebrations. To uphold these traditions, people knew which flowers and grasses to gather and where to find them in abundance. Familiarity also derived from the need to harvest and use plants for medicinal purposes, common skills that vanished as physicians gained insights into the chemicals in plants that had curative properties. Other uses for wild flowers, such as weaving baskets from the pink shore flower, thrift, also died out as man-made fibres became available. An article in the *Miami News* from 1952 ran with the title 'Old-fashioned May Baskets to Enjoy Revival This Year'. This widespread tradition of giving baskets brimming with wild flowers to deserving neighbours across America found its replacement in the mass production of cards. 'May baskets used to take much labor and skill in the making,' the reporter remarked, 'but thanks to the designers of one greeting card manufacturer, the chore is now simplified.'

Several thousand species and varieties of plants became extinct across the world after the onset of industrialization. On 4 May 1913, the *New York Times* ran an article under the headline 'Going, Going, Almost Gone! Our Wild Flowers'. 'Our wild flowers are vanishing,' argued Elizabeth Britton, founder of the Wild Flower Preservation Society at New York's Botanical Gardens. 'With each succeeding season, when the violet, the hepatica, and the Spring beauty raise their fragile heads, those who love them note with sadness how fast their numbers are diminishing.'

My grandmother keened over the blunted country places of her childhood. To her imagination, a more harmonious balance between humans and the rest of nature once existed. This symmetry of people's needs and natural processes, she believed, shaped a world more vibrant, bountiful and charming.

At the back of her *Book of Memories*, my grandmother transcribed dozens of poems and rhymes in her spidery old-fashioned handwriting; most were inspired by the Second World War or a nostalgic perception of life before those catastrophic years. The word 'remember' must have recurred in nearly half of the poems that caught her eye, such as Thomas Hood's 'I Remember, I Remember'. Many others were about wild flowers. 'The roses red and white,' she whispered to me, 'the violets and the lily-cups.' Another of her favourites was T. S. Eliot's *The Weeping Girl*:

> *Stand on the highest pavement of the stair –*
> *Lean on a garden urn –*
> *Weave, weave the sunlight in your hair –*
> *Clasp your flowers to you with a pained surprise . . .*

She had learned them all off by heart. So many poems. She repeated them to me often, especially at night-time, when I

Stream, Gailey Mill, Kingsclere 3703

was staying with her as a little girl, and we would lie in the darkness under the eaves of the house, the soft abrasion of air passing over the gaps where her teeth used to be. 'Remember and foreknow,' she echoed, 'memories, prophecies, the song the ploughman sings, the simple dream of place.'

My grandmother belonged to a generation that no longer depended on the exploitation of nature for their survival, yet they were not so far removed from it as to be entirely deprived of the skills and knowledge of their forebears. I believe my grandmother's generation still recognized the past urgency for naming and distinguishing natural resources like wildflowers. As such, those with sufficient knowledge left to them were alive to their disappearance from the landscape.

For my generation, while the use of wild flowers had greatly diminished, so had any common knowledge of them. Without this grasp of the landscape, many of us couldn't know the dull spaces filled in the past by colourful sprays of wild flowers or grieve for what no longer grew at all.

The intoxicating qualities of nostalgia eclipse the harsher need to take advantage of nature, but this is a relentless process. Once the bonds of employment overlie the Earth, the beneficial effect of people on the natural landscape and its wildlife depends in large part on their attitude, whether sympathetic and cautious, cavalier or greedy. Henry David Thoreau, a member of America's transcendentalist movement which attempted to narrow the chasm between civilized man and the landscape, wrote *Walden* in 1845. It deplored the destruction of the natural world and the acts of violence that exterminated any unwanted life form as the farmers of his country were 'levelling whole ranks of one species, and sedulously cultivating another'. Thoreau lamented the progress of agriculture and the loss of rural tradition in the lives of his countrymen. As he saw it, farming had steadily shifted from something homespun and hallowed to something immense, possessive and careless.

'We have no festival, nor procession, nor ceremony,' he wrote, 'by which the farmer expresses a sense of the sacredness of his calling, or is reminded of its sacred origin. By avarice and selfishness, and a groveling habit, from which none of us is free, of regarding the soil as property ... the landscape is deformed, husbandry is degraded with us, and the farmer leads the meanest of lives. He knows Nature but as a robber.'

I began to turn over in my mind the cultural history of our exploitation of nature. In Ancient Greece, Socrates and Euthydemus debated the evidence that the natural world was created for the sake of humankind. Socrates contemplated the comforting assistance of the sun but recognized that its rays bless all life and not only humans. As far as he could discern, all species basked in sunlight. This gave way to a far more appealing idea. What if all these other living things under the sun were themselves born and bred solely for the furtherance of people? After all, no other living creature derived so much from sheep, goats, horses and cattle. Translating Greek philosophies under the growing empire of Rome, Cicero perpetuated this misapprehension. In *On the Nature of Gods*, he offered domesticated plants as proof of the superior rights of humans. 'Their most rich and fertile fruits are of no use at all to beasts. Beasts have no knowledge of sowing, cultivating, of reaping and bringing in the harvest at the proper time,' he wrote. 'One must admit that the things I have been talking about are provided only for the sake of those who use them.'

These thinkers confused our ability to alter natural conditions with proof that the very purpose of nature was to minister to our cause. While the late origins of agriculture in human history were unknown, societies' utilization of landscapes and life forms was confounded with the image of a world ready-made for the human species. The spread of religions influenced by these philosophies ensured that the Greek thinkers' premise influenced the way millions of people considered their relationship to the natural world for centuries to come. In the Bible's description of creation (*Genesis*, Chapter 1, verse 26), God made mankind to have dominion over all other life on Earth. By the time Thomas Aquinas was writing in the thirteenth century, the ability to knowingly bring nature into service was thoroughly muddled with the presumption that the purpose of nature was to provide for humanity. Aquinas's chapter 'That other creatures are ruled by God by means of intellectual creatures' presented a presumed hierarchy of life forms from God to humans to all diversity beneath humans, as the ordered design of dominion. In the absence of evidence to the contrary, the singularity of the human mind and skills became justification for humanity's triumph over all other life. Since only we could rationalize the differences between species and bring some into service, it was clearly a part of our function as a creature to govern nature. In over a thousand years of cross-cultural musings, ignorance of the length of human history allowed agriculture to appear as an innate gift rather than the late cultural development that it was revealed to be by discoveries of early human ancestors during the nineteenth and twentieth centuries.

Time and again, the idea recurred in influential theories of the natural world and providence. In the seventeenth century, John Locke gathered these ancient speculations and tenets into a straightforward and satisfying belief that agriculture brought dominion in the original sense of the right to govern or control a large expanse. 'Subduing or cultivating the earth,' Locke wrote in his *Second Treatise of Government*, 'and having dominion, we see are joined together.' The divine right to farm led directly to humanity's proprietary rights over nature. As God and the rational powers of the human mind allowed people to subdue the Earth, Locke proposed that, in controlling, tilling or sowing the land, a person mingled nature with his own labour, giving rise to landscape as individual property.

Agricultural societies influenced by such ideas confidently assumed that farming the landscape was an act of duty and worship that called for the strict subjugation of all other life. As the American agriculturalist Jared Eliot wrote in 1760 in *The Tilling of the Land*, everything from shepherd's purse to the common mallow became an obsession of farmers for their perceived ability to 'exhaust the land, hinder and damnify the Crop'. These attitudes towards nature emerged from the severe privations of sustaining oneself from the land. But, as more intensive agriculture succeeded in filling the stomachs of all but the poorest, some began to reflect on the losses to the lushness and farrago of life.

As wildlife disappeared, many people began to romanticize what former societies despised. These compassionate values emerged because the widespread extinction of wildlife pricked the conscience of those beginning to recognize the magnitude of the losses. As William Hornaday, author of *The American Natural History*, wrote in 1897, 'It takes millions of years to produce the beautiful and wonderful varieties of animals which we are so rapidly exterminating. Unless we can create a sentiment which will check this slaughter, and devise laws for those who do not respect this sentiment . . . man will be practically the sole survivor of a great world of life.'

What appealed to me more than anything else about working in the cabin at Ding Dong was the chance to feel as if I was alone. Although there were several cottages and farms along the roadside, I rarely saw or heard anyone, unless it was a bright day. The rugged plenty of the moors stretched out as if for me. Most of all, I waited for afternoons when rains rose up from the seas below like swarms of insects, sweeping across the headland in sudden, impulsive flights. I have never felt confident walking in lonely places; there has always

been the fear of coming across danger, however unlikely –
the frustrating vulnerability experienced by women striking
out into the world. But when these afternoons came, I felt I
could venture outside, off the beaten track, for the fickle rains
seemed to keep everyone else indoors.

As well as slipping outside during the daylight hours, I
began to cross the moors at the boundaries of night and day,
when the light seemed to flock upwards or fall back, as if
weary, into the earth. As it dimmed, the world became no
more than sounds and contours to which only history and
familiarity gave form. Shadows increased in density as if the
transfusion of the oncoming night flowed through them,
while living things shrank to a cry, a pair of eyes. But if the
moonlight slit the enchantment for a moment, I might stum-
ble across a hedgehog, stock-still, as if turned to stone by my
gaze. And then it would suddenly shoot off into the darkness
again, almost unnaturally fast, released by the wizardry of a
cloud across the moon.

There seemed to be more life during these passages to and
from darkness. As night fell, anonymous birds sang out with a
remarkable and renewed vigour. Like a child's dream of toys
coming to life when the light is put out at bedtime, these
creatures of the heaths revived at the close of the day. I once
came across a fox, both of us struck for a moment as if caught
out in different acts of subterfuge. I took a torch from my
pocket and shone it over the fox, glimpsing the lava of its eyes
before it sprang off into the night.

The nocturnal habits of these animals seemed both shrewd

and pitiable. They stole their livelihoods when human dominance was suspended. Their searching eyes, their responsive whiskers and hairs, their sense of smell, all honed in the dead of night through the generations. I, too, felt I was stalking through the night to escape the order of human existence. Yet I had none of the mastery of these night beasts. Quite the contrary. My senses and skills couldn't winnow out the perils from the harmless music of the darkness or take me even a few steps forward without almost frantic effort. It took all my powers of concentration to navigate to the ruins of the mine. And once there, even though I was amazed and enticed by the dynamism of the wreckage, I could dare myself to stay for only the shortest time before anxiety forced me to sprint back towards the lights of the cabin.

I took guilty satisfaction in the absence of other people during my solitary walks. It forced me to admit that some disappearances are attractive. As the unhurried measure of the elements shaped the moor in new ways, slowly disassembling the landscape, the lack of others provided space for me to re-imagine the world. Turning from pathways defined by strangers travelling in the same direction, I sought areas unmarked by fence or stile, without any tangible sign of possession. The emptiness and solitude of a landscape to which I had no greater rights than any other person gave me an intoxicating sense of personal freedom, released from the competitive tensions underlying the crowded human domain. Yet I recognized that my presence reduced the chances of others enjoying the same pleasure. It was the uncomfortable acquaintance with the impulse to seize the world for one's own ends, often in full understanding of the unhappy consequences for others seeking in parallel.

I was thinking through these ideas on one particular walk on the moors. It was an unusually cold day. As I trod on the bracken, it hissed with frost. Rain had not yet begun to fall and the clouds were plump and dingy as old plums. A crow was perched on the stone trough, its head to one side, staring as if puzzled at its reflection in the water. Occasionally, it would peck, worrying its likeness into a hundred pieces, before pausing as the outline of its own figure returned. Then suddenly it flew off, bored by self-knowledge. I carried on in the direction of the ruined farmhouse. From a distance the broken windows appeared to admit no daylight. Night seemed to be squatting there, dense and obstinate. Looking at the farm, I wondered if we could fix our sights on some perfect moment in the past when agricultural traditions didn't press too heavily on the landscape. A time when there was space for people, too, to feel inspired and alive.

Around two million years ago, our species numbered perhaps four or five thousand and our ancestors' effect on the natural world was vanishingly small. Move through time to around forty thousand years ago, during the Upper Palaeolithic, and the human population was on the rise and hunters began to be implicated in the extinctions of the large mammal populations of several continents. Around one thousand years ago, the world population was probably just under 350 million. By the onset of the Industrial Revolution in the eighteenth century, the figure was about 700 million.

It was in the early days of industrialization that Reverend Thomas Malthus presented his views on the potential of population growth to bring about the demise of human beings. Malthus argued against the ongoing perfectibility of most societies. People had reproductive urges that they couldn't overcome, giving rise to conditions that sometimes impeded progress. The extinction of the passion between the sexes, he conceded wryly, seemed impossible to achieve. But one of Malthus's contemporaries, William Godwin, responded fervidly with a vision of the world wholly cultivated and improved by the presence of a multitude of people, all of whom could aspire to progress and wisdom. Some of Godwin's hopes came to fruition in the stupendous achievements of industrial processes and technologies. But he did not foresee their disastrous ramifications, which allowed people to evade some of the restrictive laws of nature.

His confidence derived from the works of Francis Bacon, who sought to vindicate the dominion of humankind above all else. Writing in 1620, Bacon argued vehemently for the merits of capitalizing on nature by means of technological innovation. From his perspective, the point of understanding the natural world was to help humans to gain ascendency over the landscape. In the first book of *New Organon*, Bacon identified three kinds of ambition: the ambition to extend an individual's powers, and to further the power of one's country; 'But if a man endeavour to establish and extend the power and dominion of the human race itself over the universe,' he concluded, 'his ambition . . . is without doubt both a more wholesome and a more noble thing than the other two.'

It was a thrilling endorsement of humanity's technological prowess and the justification of its use for the primacy of human societies across the Earth. This ideology underpinned the escalating technological feats that affected every facet of existence, improving sanitation, housing and health care in many countries and societies, extending people's lifespans, and, perhaps most pertinently, enhancing methods of food production by orders of magnitude. One hundred years after the onset of industrialization, the world population grew by 100 per cent. By the start of the twenty-first century, it had increased by a further 400 per cent, a colossal surge that led to the daily threat to the survival of other species on the planet.

2

Tin

In the mounting daylight, a soft morning mist turned the view across the moor upside down like an hourglass, flinging the sea to the heavens. Spring was brightening into summer and, for several weeks now, more and more walkers and tourists had been visiting the ruined engine house at Ding Dong. They startled the birds from their roosts, taking photographs and staring at the structure in silent beguilement. I kept returning to the question of what attracted them. Why were the gutted remains of an extinct industry such a captivating shape on the horizon?

I began to see that the idea of loss was riveting to the imagination, like shrapnel lodging in the mind as a permanent ache. I began to believe, too, that this was an involuntary response to losses discovered both in the external world and in the internal life of an individual. But why did people suffer these aching emotions, if not for some purpose? Were they the consequence of some other aspect of sentience or were they in themselves rewarding feelings?

While studying the early English language at university, I came across an ancient English word for the fascination experienced by someone looking at a ruin: *dustsceawung*. A word of startling precision for which no modern equivalent exists. A kind of daydream of dust, a pondering of that which has been lost: dust-*seeing*, dust-*chewing*, dust-*cheering*. The daydream of a mind strung between past and present. The word was preserved in a tenth-century anthology of poems safeguarded for a millennium in the halls of Exeter Cathedral. One of the poems in the collection, 'The Ruin', was itself ruined by an unexplained fire, some of its beauty and meaning intensified by flames. It was composed by the great-great-grandchild of one of the warriors from the Germanic tribes who crossed the seas to profit from the wreck of the Roman Empire. The poet found uncanny solace and beauty in the lichen-shadowed walls and cemented garrets glistening with frost. *Wrætlic is þes wealstān*, he gasped, 'the stonework is wondrous', *wyrde gebræcon* . . . 'broken by fate', *burgstede burston* . . . 'a city smashed', *brosnað enta geweorc* . . . 'spoiling the work of giants'.

The poet and his society were struck by the wreckage of the empire, sunk as they were in its lost glories. The conspicuousness of Rome's obsolete architecture across the lands it had occupied demanded imaginative attention. Some of the poets of the Germanic tribes peopling Britain began composing hymns to those they called '*wreacca*', the origin of the modern term 'wretch'. Wretches were wanderers singing at the threshold of place and tradition. One of these hymns, entitled 'The Wanderer', was an anonymous poem set down during the ninth century, though doubtless sung earlier in pliant forms. The wanderer in this narrative finds himself alone, unsettled and drifting. Memories of the dead and fantasies of his friends swirl across the acres of his exile like sea frets along cliff tops. The language he uses is elliptical, memories of his comrades and kinsmen melding into the silhouettes of seabirds, the souls of men in Nordic cultures. One line of the poem, which might mean 'sailors' ghosts bring few familiar songs', begins with the word *flēotendra*, which in early English meant seabirds as well as sailors. The keening of the gulls is little recompense for the cheering songs he shared with his companions. In this state of unreality, his grief-struck mind dreams of a ruined city across which all voices fall silent: there are no survivors.

> *Walls stand, wind-beaten,*
> *Hung with hoar-frost; ruined habitations,*
> *Human laughter is not heard about it,*
> *And idle stand these old giant-works.*

The wanderer in the poem both yearns and suffers as he travels further and further from the real world of his memories. The greater the distance between him and this familiar life, the more acute his fear that the memories will distort and fail to find any continuity with physical truth.

The word closest to *dustsceawung* I could muster from modern English was 'nostalgia'. In the seventeenth century, the Swiss physician Johannes Hofer combined the Indo-European root *nes*, found in the Greek *nostos*, meaning the return to light or life, with the Greek term for sorrow, as he sought to describe the disease he witnessed among soldiers far from their native lands. Recognized as the pain individuals endured as they yearned to return to their home, nostalgia was, to Hofer, 'a continuous vibration of animal spirits through fibres of the middle brain in which the impressed traces of the idea of the fatherland still cling'. To begin with, people understood nostalgia as the comprehensible, physical condition of a mind and body in want of its home. But as use of the word and idea increased, other notions began to cluster around it; nostalgia became a sensation that might sway between the pain of the medical disease and the gladdening desire for a prior life.

A lingering study of something absent or lost, a baffling compound of emotions both poignant and pleasurable, nostalgia was often dismissed as sentimentality, a pointless longing for a vanished place or time. But I began to consider whether the sight of something extinct or on the brink of extinction provoked a more constructive sense of yearning for all that threatened to tumble beyond reach.

The whole landscape of Cornwall expressed dereliction. Once the strings of white pebbledash houses petered out into the occasional cottage, the landscape became one of raddled hedges and dark knuckles of granite. Something insubordinate remained, perhaps fostered by the nearness of the great Atlantic Ocean, a sense that the elements merely tolerated human presence, biding their time. The ruins of the mining industry were preserved across this landscape in tallowy grasses, descending unhurriedly into their graves. As an outsider to the region, who had never known the active industry, I took from this broken vision only pleasure and interest.

On my way home one evening from the nearby city of Truro, I glimpsed some huge ruins from the roadside at Carnkie. Out of curiosity, I stopped and decided to seek them out. The path that led to them was muddy and overhung by an ancient hedgerow. As I walked, my footsteps startled partridges into the sky. It had been a warm afternoon, and the setting sun brindled the leaves and grasses with what remained of its light. Two young, scruffy boys, who stared at me with hostile wariness, were playing football in the dusty car park. I listened to their whoops and jeers fade as the backstage harmonies of birdsong and rustling trees steadily overcame them. Then, suddenly, the abandoned mine came into view. The hedges and the ivy had made a secret of it. Excited, as if I had discovered the treasures of an ancient civilization, I hauled myself up through the half-collapsed arc of a large window and studied the structure and the views of the country beyond. At first, an electrified hush surrounded me; then I heard movement and had a moment's fear of being alone. A man was riding up the lane on an old, rusty bicycle, his head down. He was somewhere in his sixties, perhaps, or older, wearing a frayed and faded navy-blue jumper. As I moved, he looked up and slowed. He let his bicycle fall against the hedgerow but did not enter. He only stared at the mine from the lane and then carried on his journey. Something in his gaze convinced me he was a former miner.

In the coming days, I found myself unable to shake the man from my thoughts. I continued to walk the moors every day, but spent more and more time just sitting against the mellowed, summery stone of Ding Dong mine. The pinkish bricks of the chimney had eased from their rigid framework to create a more leisurely outline. Every crevice and plane clutched a handful of grasses and shrubs, offering up design to the anarchy of nature.

I thought about the explanation of memory given by Thomas Hobbes in *Leviathan*. Sometimes a man seeks what he has lost, Hobbes concluded, obsessively returning in his mind, searching through places and times for the moment when he still possessed it. And then he pores over these memories to establish what caused the loss in the first place.

If people experienced nostalgic feelings when looking at something extinct, was this a natural encouragement to seek the causes of its demise? Whereas the old rural rhythms may have been in harmony with nature, the obvious destructiveness of mining made its loss harder to lament. Some four millennia ago, the environmental effects of the tin industry were already prodigious, as entire lengths of the country were deforested. Yet its remnants were objects of beauty, utterly captivating. And I felt they had to be compelling for a reason beyond any romantic ideas about ruins. I found an archive of films about mining, made by the BBC, spanning the decades between the 1930s and the 1990s. One film opened with two Cornish miners and one of their wives seated around the kitchen table. It was 1985.

'Look at the papers today,' one of them said. 'Fifteen thousand miners on the dole. Going to have to think about going away.'

His wife glanced up at him. 'You know yourself. What are we going to sell these houses for?'

'Next to nothing. We need the backing of the government. It'll cost 'em more to keep us on the dole than to have us in work. They don't give a bugger for we down here. They're lining their pockets, they don't mind about we. All they want for Cornwall is a holiday centre and they're going to get it.'

In the 1980s, the managers of the Geevor tin mine requested a subsidy of twenty million pounds from Margaret Thatcher's Conservative government to overcome the slump and keep the mine open. The request was refused.

When the industry died, the younger men turned their hands to something else, but for those for whom mining had been a lifetime of growth and selfhood, the loss was harrowing. After the collapses of the 1980s, many British miners didn't secure further work and remained dependent on government welfare until their death. The last of the BBC's tin-mining films, made in 1986, summed up the grief experienced by those forced out of an industry they had worked in all their lives. One man said quietly as he received his redundancy and eviction notices, 'I never really wanted to do anything else. My people were miners, or engineers connected with the mining. It's a different world underground.'

Geevor was one of only four mines that survived into the 1980s. When its demise came, it was sudden. 'Four short weeks ago this was a mine,' said the deputy chairman, Ken Gilbert. 'Now look at it. It has become in those four short weeks almost a ghost of what it was before.' Days later, he went around the village where over half of the inhabitants were employed by the mine and gave eviction notice to those in mine cottages.

The old miner said, 'Mr Gilbert came to the door. He handed me the letter and he stayed here with me as I opened it and read it. And he was sorry himself, you could see that . . . I still haven't got over it. But now I feel that I want to get out of the houses anyway. There's no work, the village is dead, so it's just as well I get out of it.'

These people grieved more acutely than might be accounted for by economic hunger or pride in one's position.

Their resourcefulness drew them back to the former vitality of mining, hankering after thousands of years of experiment and accomplishment.

Tin ore typically formed in matrices of quartz and schorl, with aureolas of slate and granite, sometimes many feet thick. Above this were pallets of sand, silt and shells, needled by fragments of trees, stag antlers, tines of deer, skulls. In the early part of the fourth century BCE, the Greek mariner Pytheas anchored in West Penwith on his circumnavigation of the British Isles. The inhabitants, he said, were busy preparing tin, hammering the metal and shifting it to an island off the coast called Ictis, where merchants sailed to trade. Having begun in the Bronze Age, active tin mining defined the landscape of Cornwall until the end of the twentieth century, one of the oldest continually operating industrial areas in the world. Over the thousands of years that the Cornish worked in mining, a culture established itself as the heritable knowledge of the generations. For centuries, the Cornish word for mine was *bal*, a corruption of the old Cornish word for digging by hand, when digging was still the basic, back-breaking method of mining. In the mid fifteenth century, miners realized that they could go further underground, tunnelling into cliffs; it was gruelling work that involved heating the rock and throwing cold water on it until it cracked and revealed the ore. From the seventeenth century onwards, they used gunpowder to break rock, while the combined force of horses and men lifted out the water that threatened to drown the mines. In the eighteenth century, Thomas Newcomen and James Watt developed beam engines to pump the water.

The Watt steam engine, a hugely successful design, was patented and gained a monopoly. Behind such inventions were small-scale improvements to design and efficiency spurred on by competing individuals and businesses, and often by neighbouring workers who shared information with one another. The 'adventurers' were Cornish miners who studiously battled to improve the design of the engines that pumped water from mines. Towards the end of the eighteenth century, the adventurers rebelled against the monopoly on Watt's steam engine, installing modifications and new models based on their own experiments and ideas. They published a monthly journal, *Lean's Engine Reporter*, which detailed their operation tests and technical findings, diffusing their knowledge throughout the region. The Cornish mining engineer Richard Trevithick was chief among those who benefited from this exploratory atmosphere. Trevithick elaborated the earlier models through the use of high-pressure steam, which led directly to the world's first road locomotive powered by steam. He nicknamed it the 'Puffing Devil' as it made its virgin voyage across the hills of West Penwith.

At the outset of a mining operation, adventurers divided up shares in the mine to reflect the costs of the venture. Each quarter, the mine accounts were perused. Losses invariably urged on further inventions. As one of Cornwall's leading adventurers wrote in 1830, 'We avail ourselves of the assistance of many; and the great scale upon which we have to experiment makes the result most interesting to us.' Trevithick shared his steam technology, enabling the industrial extraction of minerals and metals all around the world and contributing to the rise of the railways. Through these advances, mining went deeper than ever before, 330 fathoms into the prehistoric granite of Cornwall.

During the heyday of mining, miners were old by their forties. In 1837, a survey found that their average lifespan was thirty-one years. Predictably, they were superstitious, unwilling to work on Midsummer's Eve or Day, cautious not to whistle underground. The dim glow by which they worked flickered from a candle stuck to the gallery wall with clay. A shift was eight or nine hours, and progress was agonizingly slow, a few inches a day. The men sang to give solace, their voices resounding preternaturally through the steamy tunnels, 'Ho! Brothers, ho! How ye do thunder! The earth grows young again in you.' Ascending to the surface was 'coming to grass'. Some mines were so deep it took an hour for the miners to emerge. Once there, they gasped like stupefied fish out of their element, puffing out sighs of dark, smoky breath. Children as young as eight or nine worked in the mines, dressing the ore alongside women. Self-conscious girls wrapped their ankles in heavy thongs of wool. A commission in 1842 interviewed dozens of children, who complained of shortness of breath and backache. Few of them could read or write. Martha Williams, whose father had 'been dead this brave while', began working at Trethellan mine when she was ten. She rose at four and walked five miles a day to and from her workplace.

Like farming, mining was a hard life, but it enhanced an

intuitive perception of the landscape. At a mining museum near Land's End, I read about the signs that allowed miners to establish the presence of the ores simply by scanning the ground with their eyes. They knew the plants that thrived as their rootstocks pressed like tongues into soils enriched by minerals, as well as those that shrivelled in the presence of the metals. They could also sip from a stream and use their taste buds to riddle the waters for the subtleties that indicated the presence of the precious minerals. Or else catch a hint of pyrite in a heavy intake of breath. Acutely absorbed by their landscape, the miners gained their knowledge of the ancient rocks that underlay their country and of the natural conditions and creatures that ornamented the surface.

In June 1797 John Frere, an antiquary from Norfolk, wrote to the Society of Antiquaries with an account of some flint weapons he had discovered at Hoxne in the neighbouring county of Suffolk. These objects of curiosity, he said, were evidently weapons, fabricated by those who did not then have the use of metals. In the same layer of the earth, he found 'some extraordinary bones, particularly a jaw-bone of enormous size of some unknown animal'. In the letter, he made a speculation that was radical for his era. He mooted the possibility that these stone curiosities and extinct creatures both belonged to an extremely remote time, 'even beyond that of the present world'. In other words, humans might have had a considerably longer and more perplexing history than the Bible implied.

It was several decades before hundreds of other finds, particularly from France, persuaded Frere's contemporaries that an earlier stage of humanity evolved some of the first stone-tool industries. This culture of tools became known as 'Acheulian', from the area near Saint Acheul in France where archaeologists unearthed some of the culture's artefacts. These early people clasped the large, often oval hand axes long before the appearance of modern humans.

What interested me about these discoveries was the associated evidence that emerged decades later. As archaeologists excavated more fossils, it appeared that during some of the first incursions into unknown and uninhabited landscapes, some groups initially lost the tradition of manufacturing tools. The fossil evidence suggested that it was only when further populations spread out from the original sites of the innovations in Africa, tens of thousands of years later, that knowledge of the technology was renewed. Perhaps when the hominids found themselves in utterly different rocky surroundings, with none of the familiar stones to rouse interest, the Acheulian industry simply died out after one generation, even though it was of great consequence to their survival. Some groups struggling to hunt without tools were sure to become extinct.

Eugene Dubois was an excitable, persistent young Dutchman who abandoned his profession as a doctor to pursue his fixation with the ancient ancestry of humans. In 1891, at Trinil in Java, he found the skull of a hominid he called 'an erect ape-man', the first discovery of *Homo erectus*. His analysis was dismissed by contemporaries, which frustrated and hardened him. In a fit of pique, Dubois hid the bones in his house and refused to discuss them for over twenty years.

In 1907, in the river sands at Heidelberg in Germany, the lower jaw of another hominid was excavated, *Homo heidelbergensis*. A decade later, Johan Andersson, a mining advisor to the Chinese government, met a friend who recounted the superstitions about a place outside Beijing called 'Chicken Bone Hill', where the red clay was riddled with large numbers of bird bones. Andersson, who was fascinated by fossils, went to investigate, stripping from the ground a rich haul of animal bones – wild ox, wild pig, rhinoceros, horse, tiger, hyena, mammoth. His searches eventually led to the site of Chou K'ou Tien cave. There he and his companion, Austrian palaeontologist Otto Zdansky, found pieces of quartz, which had 'such sharp edges that they might very well have been used as cutting tools'. Noticing quartz in the roof of the cave, Andersson concluded that such tools would have been picked out by the users rather than fashioned by them. Andersson became convinced he would find hominid remains. In 1927, Zdansky dug out some molars, which he cautiously labelled *Homo sp?*. Soon after, he and Andersson discovered two crania. The cranial bones were markedly thick, similar to recent finds in England, but the leaden arches of the eye sockets and the low forehead bore similarities to Dubois' Javan fragment.

After the Japanese invasion of northern China, excavations at the cave ceased. The hill was fortified, the field laboratory smashed. Work only resumed in 1949, when the chaos of war had stilled enough to allow researchers to clear the rubble and rebuild the infrastructure. Thereafter, five teeth were excavated, alongside fossil animal bones and stone artefacts. Professor Wu Ju-Kang and Mr Chia Lan-po from the Laboratory of Vertebrate Palaeontology in Beijing ran studies on them, confirming their antiquity. On the basis of this evidence, in 1953, the site was transformed into a park for tourists. In the first year if its opening alone, twenty thousand people visited. Eventually the fossilized remains of over forty hominids were discovered, men, women and children, who lived at the caves over half a million years earlier.

None of these early humans possessed hand axes of the Acheulian kind. Possibly they had not yet been invented. Or, quite probably, these people had lost knowledge of the technologies some time during their migrations. Although there were only scraps of evidence of early people from sites scattered far apart, one evident pattern was that the different bands of hominids that attempted colonization outside Africa ultimately failed in their endeavours, fading away into extinction. At Gran Dolina cave in Spain's Atapuerca region, several hundred artefacts were embedded alongside the fossils of six hominids. There were no hand axes. The colonization efforts of these hominids, tagged as *Homo antecessor*, came to nothing. The earliest confirmed sites of human ancestors in Europe date from around 800,000 years ago; archaeologists speculate that this hints at the challenging glacial conditions these people faced before then.

The head of *Homo heidelbergensis*, with its large, prognathic face, massive, chinless jaw and anvil forehead, progressively developed to house a much enlarged brain. Whereas the early Acheulian tools were thick, rutted, uneven hand axes, the later tools were more evenly shaped and skilfully crafted; this greater technological prowess possibly allowed for more successful settlement of other landscapes. It's thought that the crafted tools of the late Acheulian were made possible by the cognitive leap of rotating the tool in the mind prior to sculpting it. No one is able to say definitively, but the innovations of the broadminded *Homo heidelbergensis* were probably essential to their greater success in migrating through Asia and Europe around half a million years ago, as they expanded out of Africa again. Still, these intermediate hominids all eventually died out too, possibly due to competition from the more successful ancestors of modern humans.

Evidence from Ethiopia shows that modern humans evolved from African populations around 120,000 years ago. The slightly earlier *Homo sapiens idaltu* were probably users of Acheulian methods, but by the time they evolved, *Homo sapiens sapiens* had outgrown Acheulian technologies. Clutching much improved stone tools, they succeeded in colonizing the world to become the direct ancestors of everyone alive today.

These discoveries suggested to me the importance of holding on to industrial knowledge for our survival as a species, and perhaps why the urge to save or even increase the destructive potential of some technologies overcame their disadvantages.

Mining for metals significantly altered landscapes, un-burying potent elements from great depths that embittered soils – cadmium, arsenic, zinc, iron. The minerals that miners crushed and smelted into workable forms left traces in the earth, in rivers and streams, and in the air. Pollution from mine waters stunted the populations of fish and amphibians, especially in upland streams where the animals formerly bred. This, in its turn, lessened the feeding opportunities for birds and mammals further down the rivers. A recent survey of the soil around mines found indicators of metal pollution dating from the mid Neolithic, over four thousand years ago, rising distinctly during Roman occupation, and again during the onset of industrialization. I can still remember clearly from my childhood the pollution of the estuary of the River Fal in Cornwall. The disaster unfolded in the winter of 1992 as mil-lions of gallons of heavily contaminated water flooded out of the abandoned tin mine at Wheal Jane.

It was not only animals that were in danger of extinction. The swift removal of a natural resource that took millions of years to form would inescapably lead to the exhaustion of supplies. An edition of *Time* magazine from November 1926 highlighted the need for tin in the United States, for every-thing from solders, printing callicoes and collapsible tubes to bearing metals and tin foil for wrappings. 'For these uses,' the author noted, 'this country last year imported 50 per cent of the total world supply of tin,' ending with a cautionary state-ment, 'There is a world shortage of tin.'

One afternoon, while I was walking the coastal path around

West Penwith, I came across an old miner and his wife. It had rained all morning and the moistened ground released a fertile perfume of turf and cloves. The small flowers of the heather softly bruised the headland with colour. In among the foliage, I spotted a yellow triangular hazard sign, the black silhouette of a man falling headfirst into the opened jaws of the earth. BEWARE MINE SHAFTS! KEEP TO FOOTPATH. I saw two figures on a bench a little way ahead. 'Hello,' I called, jovially. They were perhaps in their late sixties, both beginning old age. The woman had a round, welcoming face. She began talking to me; her voice was lively, girlish even, asking me if I was a tourist. I told her of my interest in the mines, shouting over the bluster of the wind. As we spoke, the man beside her remained still, his eyes trained on the sea. He seemed shy, his stance eloquent in its self-containment.

'Well, you ought to ask my husband about that,' the woman continued, nudging him. 'He was a miner.'

'Oh, really?' I said, directing the words to him.

He glanced up at me, slightly lifting the lines that mazed his forehead. His eyes were a pale, startling blue, the sky seen through ice.

'Yes,' he said, nothing more.

'What kind of mining?'

'Tin.'

'When did you retire?' I asked him.

'More than twenty years now.'

'Do you mind that the industry is finished?'

He shrugged. 'I'm out of it now. But it's a shame to see it go.'

The problem, the miner said, was that there was both too much tin and not enough. A pattern established itself, he explained; with the emergence of a new competitor, such as the open-cast operations in the Far East, localized slumps occurred elsewhere and the mining tradition of the labour in places like Cornwall came under threat. When profits collapsed, mines were flooded and left to die, the pumps stripped out and sold to foreign operations. Miners found themselves employing their skills in new continents, as workers or managers of mines for lead, copper, silver and gold in regions as far-flung as Colorado, South Dakota, Ontario, Wellington or the Cape of Good Hope. Cornish surnames like Lanyon, Jelbert, Furze, Tregidga appeared in the censuses of towns like Moonta in South Australia and Pachuca in Mexico.

In a letter to his sister, Caroline, composed in Valparaiso in October 1834, Charles Darwin described his visit to a Cornish miner working in a ravine in the Andes. From the first hints of trouble in the early nineteenth century until the disruptive force of the First World War, Cornwall lost a quarter of its male population, predominantly through the pursuit of mining opportunities overseas.

Orient Mine, Trespuntas, Chile, South America,
June 26, 1852
Dear Father, Mother, Brothers, and Sisters,
　　Our mine is up in the desert of Chili. There is nothing here but what is brought on mules backs . . .

By the mid twentieth century, only a stubborn remainder of tin mines hoped to stand proof against the vagaries of the global markets and the bitter end of a finite resource. In 1998 South Crofty, the last Cornish mine, was closed. In 2010 it reported findings of commercially viable quantities of gold and thousands of tonnes' worth of the rare metal iridium, used in satellite navigation systems, touch-screen technology, televisions and mobile telephones. Mining began again in January 2011.

I sat on the grass of Burnt Downs by Greenburrow shaft, erected in 1865 to house a forty-inch cylinder pumping engine. In the eastern distance were Ishmael's Whim engine house, Hard Shaft and East Killiow Shaft. Across the moor, close to the hamlet of Tredinneck, was the former count house, the eastern boundary of the tramway. At the end of that decade, Ding Dong had expanded its workings, employing over two hundred workers. But it soon went the way of the others, closing for the last time during the First World War.

The magnetism of the industrial rubble engaged onlookers with the history of those gathering enough sense of nature to successfully exploit it, material remnants in which past experience survived. Once the last tin mine in Cornwall had closed, a campaign sprang up almost immediately to preserve the wrecked industry for tourism. The inspiring quality of ruins left people susceptible to puzzling engagements with these places. Counsel from these ruins could easily dissipate in the unrelated surroundings of the visitor's origins.

If ruins were not to be of service to those living in the area, I preferred the image of nature laying hold instead, asserting its prior rights of ownership over abandoned areas, like a monarchy in exile making its triumphant and poignant journey home.

In the natural order of things, the scourge of the wind and rain weathered the rocks, but miners had dug them up, fired them, pounded them to dust, poured toxic substances into springs and streams. Over centuries, mining had created unique habitats colonized by spindly, obdurate species tolerant of poor, slightly poisonous soils. Some kinds of life, like beetles and lichens, bloomed and prospered in these habitats, especially once they could take possession of places abandoned by people. Rushes stood alert in the sunken land, while mosses and heathers sprawled across and tinged the hummocks. The extremely rare Cornish path moss, a species found only in Cornwall, clings to pathways and tracks, one of the few plants sympathetic to traces of copper in the concentrations found at ruined mines.

In the twentieth century, the snub-faced greater horseshoe bat, which flitted about the ruined mine of Ding Dong in the mauve dusk, had suffered a dramatic decline across northern Europe. The loss of traditionally grazed pasture, along with the spraying of pesticides that destroyed the insects the bat preyed on, caused populations to plummet. Its roosting sites appropriated by people, its woods razed, the bat struggled to find shielded homes. The ruined mines became its sanctuary. So, too, the dun-coloured nightjar, its round, astonished eyes inquisitive of darkness. In Britain, numbers of the birds had dropped by more than half since the 1960s. Roosting males would withstand both gale and blizzard in order to find freedom from disturbance. The broken windows and doorways of derelict mines, the lightless silence of the shafts, provided homes for peregrines, nightjars, bats. And so, one of the strongest sensations I experienced during my days at Ding Dong mine was of trespassing on a world forsaken by man and now the home of others that did not, in truth, look on my presence with favour.

3

Ghosts

An unpleasant feature of walking across the moors alone was the sense of disorientation when the winds gathered their energies and scattered the rains in every direction as if maddening them. In a short time, I would lose any certainty as to where I had come from or where I'd hoped to go, as the place itself lost its sharpness and poise. A feeling of panic quickly asserted itself and would not abate. No matter how much I told myself that I was being ridiculous, that walking in a country where there's always a road reasonably nearby meant I had access to any assistance I might need, still anxiety thrummed inside me. At these times, I chided myself for going out on the moors alone and would recall one of my favourite ballads, 'The Banks of the Lee', in which a young woman meets with death while out on the moors, waiting for the return of her loved one. 'Don't stay out too late, love, on the moorlands for me . . .' Quickening my pace, stumbling over the crackling branches of bracken and gorse, stamping the black mud of barer ground, my heartbeat remonstrated like clicking fingers in my chest. But if I could calm myself, I would begin to notice details in the landscape that I might otherwise overlook: the counselling line of a stunted tree, a sequence of hollows in the undergrowth. These idiosyncrasies acted as guiding lights – by them, I could navigate my way back to the ruined engine house.

As I walked, my interest in nostalgia and its relationship to extinction grew, because it offered me hope that it was also natural to us to be protective of nature. As such, we could take measures in our lives to increase this tendency at the expense of other selfish, hostile aspects of our character. I began to wonder if nostalgic impulses were vestiges of the kind of memory prehistoric people developed as they spread into farther reaches. These journeys and diffusions caused us to be a global and dominant species. But what kind of animal had migrations made us? What feelings had such navigations sharpened into reflex and thought? To survive, the wanderers on the borders between known places and new would explore mountains, rivers, woods and grasslands, learning to live off the land, searching for indications of how previous people sustained themselves. Intimacy with nature was critical. The movement of the grass reflecting the direction of the wind; the slim shadows of rushes setting time by the sun; patches of tawny ground marking drier steps across a marsh. Slight signs like these, among countless others, helped wanderers navigate in the wild. But such signs were only of service if people could discern and remember them.

In the early nineteenth century, the French philosopher Maine de Biran first distinguished 'mechanical memory', or habit, from other kinds of memory, in his long essay 'The Influence of Habit on the Faculty of Thinking'. He believed that some kinds of memories derived from repetition that turned into unthinking habit, while others seemed to be acts of conscious reasoning. In the 1960s, the Canadian neuroscientist Dr Brenda Milner provided empirical evidence for Maine de Biran's suppositions. Conducting a series of experiments at the Montreal Neurological Institute on patient 'H.M.', who had no ability to recall recent events, Milner was able to disentangle two different kinds of memory: 'procedural' and 'episodic' memories. Like learning to ride a bicycle, procedural memories originated in actions repeated until patterns were imprinted in the body, achieved effortlessly without conscious thought. Episodic memories were conscious experiences, blends of places, encounters and sensations. Emotions criss-crossed such memories like ivy twisted round bark.

Several million years ago, early hominids were gradually exploiting larger and larger areas. The African climate was growing drier and cooler and light began spearing through the dense canopy of the giant forests. Some ill-favoured creatures, such as the giant gelada baboon, *Theropithecus brumpti*, which once thrived under the treetops, disappeared from the fossil record around this time, unable to outlive the changes. Mammals like *Theropithecus brumpti* and the ape ancestors of humans possessed only habitual memory. But the human ancestors that survived the retreat of the forests gradually evolved memory capable of preserving singular happenings in the mind, elaborating events and objects through previous experiences, straining beyond immediate perception. They moved out from the cowling of forests and into the open country of Africa. The weather was erratic and those that survived were increasingly resourceful and inventive.

Through countless millennia, these hunters made tracks into virgin wildernesses, unexplored, exotic and dangerous. Their minds broadened and reappraised memories, focusing on objects with life-and-death significance. Explored and settled places became intriguing to people, the traces of previous habitation visible in the ruined or abandoned remains of landmarks and communities.

The habitual cognitive skills of ape ancestors were blinder, without conscious reflection. But the emotional memory that eventually emerged among modern humans was exceptionally adept at recalling and refining skills, and understanding why these were significant. They possessed a memory that saved them from extinction. It was the potential of this memory that captured my imagination, the poignant faculties of a more forgiving, versatile mind.

As I came to the end of my time in Cornwall, I escaped from my work along a quiet country road graven into the moorlands, past the ruins of Chysauster, an Iron Age village settled during the Roman occupation of Britain in the first century AD by a Celtic tribe known as the Dumnonii. Following signs to Cape Cornwall, I crossed on to an uneven track, flanked by fields in which the ruins of two engine houses stood. In the distance, I could see the ornamental brick stack of an old mine, which seemed to mark the end of land. The sea took shape beyond it, exquisitely calm as if in rapt attention.

The slumped stonework of an old chapel lay in the fields behind the stack, walkways threaded through the greenery to glassless windows widened by decay, where grazing sheep and lambs sprang in and out of the hallowed and common land. A small board with notes and a map confirmed the edifice as Helen's Oratory, one of the first chapels raised in Cornwall in the fourth century. The sun propped up the failing shape of the chapel, whose fractured roof lay sunken in light. Embossed on granite, an old, weathered cross praised the sky. Daffodils nodded penitently in the breeze. A whole landscape of vanishings surrounded the chapel's remains – not only the vestiges of mining but also mud-locked archaeology from a Bronze Age burial ground, and the banks of Kenidjack Castle, an Iron Age fort. Below, at a place named Priest's Cove, the rocks cradled an ocean from the Late Devonian era, 380 million years ago. Even the name was an anglicization of the extinct tongue of the Cornish, originally Porth Just, pronounced por'east.

Perched on a small plateau overlooking the crescent of the bay were the spoiling, storm-strong structures of abandoned fishing huts. I had that lovely feeling of potentiality inside me, alone and ranging directionless over a landscape. Further along the shore was an old fisherman's shack. Balancing and slipping on the boulders, I reached the rocky platform and entered the small, lichen-clad building. My hands were sore and bleeding from the jagged, salty rocks and I blew on them gently to cool the stings. Inside the hut, the sea sounded

nearer, echoing and multiplying in the snug space as if it were the heart of a conch. I imagined a fisherman sheltering from a storm inside these cramped quarters and the terrifying excitation of the waves. It was dark despite the splendour of the afternoon and as cool as a refrigerator.

As my eyes adjusted to the gloom of the interior, I made out names and dates etched coarsely on the stone walls. *Kelly loves John 1996.*

The fisherman's hut had become a playground for lovers.

I remembered, as a teenager, coming to West Penwith on holidays. A short walk from the centre of Penzance along the arc of the harbour took us to the old fishing town of Newlyn, the last port visited by the *Mayflower* en route to the New World. A small lane wound past the nook of the harbour with its trawlers and an old restored sailing lugger, and uphill, where a snag of wires and lightbulbs spelled s-a-v-e o-u-r f-i-s-h. As we wandered through the town, I used to stare at this and reflect on what it meant.

For hundreds of years, the only fish that Cornish fishing fleets could market in bulk was pilchard. Matthews' *Guide to St Ives*, a small town close to Penzance, published in 1884, described how shoals of pilchard began to silver the waters during October. At a wooden hut on the hills overlooking the sea, men called 'huers' watched for the approaching shoal and communicated its movements to those below by flourishing two wooden frames woven with branches. Shouting, 'Heva! Heva!', people rushed to the shore to help gather the harvest of the sea. The author explained that 'heva' was a word of great antiquity, from the old Cornish word *havas*, meaning 'found'.

Streets were often so narrow that during the pilchard season, fishermen carried their catches themselves in 'large maunds' (wicker baskets), rested on the men's shoulders using poles. They lugged them through the village to extensive cellars, where, in weeks to come, they would pack the fish in wooden barrels by candlelight. In *The History of Cornwall*, published in 1824, Fortescue Hitchins reported that although fishing was pursued with equal eagerness and skill, it was widely believed that fewer fish were caught as the years passed. He quoted from a source that claimed an average of twelve thousand hogsheads (wooden barrels) were ordinarily packed in the early eighteenth century, in the Fowey region of Cornwall – some thirty million fish. During 1816, thirty seiners were made in Fowey, yielding catches that filled ten thousand hogsheads. It was deemed a good year, in which the greater range and potential of the boats played an essential part. 'Necessity rather than choice was the principle stimulant to enterprise.' None the less, Hitchens reiterated the general view that the shoals of recent years contained 'less bodies of fish than those which visited our ancestors in former periods'.

The pilchard fishery began to decline rapidly in the second half of the nineteenth century. The *West Briton and Cornwall Advertiser* of 6 April 1852 reported that 'The zealous sean [seine] caught, on Saturday night last, in Gerrans Bay, about eighty hogsheads of pilchards. They were, however, of a very small sort, and not well adapted for the foreign market.' Sixty hogsheads were taken at Fowey. But these, too, were too small for curing. Although far more significant in populations of long-lived fish like bluefin tuna, the smaller size of the fish suggested that over-fishing had left only the young. In tandem with these changes, the modest fishing boats gave way to larger, modern vessels, powered to stay at sea for long stretches and to sail further afield. They could hold catches larger than the old boats could carry to a safe landing-place. Anecdotes from the time echo each other – the fish slowly retreated out to sea and the vessels were steadily modified in a bid to maintain profits. For centuries, a caller stood in wait on the cliffs, the first to report the size of the catch to the community. As finances were more tightly controlled, luggers and their catches were formally listed; the officers in charge would calculate the profit and Saturdays saw the doling out of each fisherman's share. The pubs unlocked their doors.

Until the 1950s, a double-ended lugger was the main boat used by Cornish fishermen. Originally, they were fairly slow-sailing boats but as fish stocks declined, shipwrights kept up by designing faster, more efficient models. These boats could haul heavier catches through rougher seas to cater for the fish markets of the cities. During the First and Second World Wars, many luggers lay idle as the men of the villages went abroad to fight. When they returned home to resume fishing, they found the waters brimming with fish. Skippers and investors made good on this temporary boon. Between the wars, motors made it possible to haul lines with six thousand hooks, three times the amount that men could haul by hand. By the 1950s, shifting to new and faster boats better enabled fishermen to keep their profits high.

Fishing was a minor industry in Britain by the close of the twentieth century, propped up by subsidies, bolstered by nostalgia for its former importance. It was much the same for fishing industries around the world. Evidence suggested that within little over a decade of switching to industrialized techniques, the diversity of species drastically contracted. The experiences of Cornish fishermen were associated with a global phenomenon. As small-scale inshore fisheries went through boom and bust, fishermen hoping to maintain their livelihoods headed further afield in search of catches, intrusions that called for better technological faculties. As a direct consequence, a disastrous situation arose whereby catches often appeared to increase due to superior fishing methods, while populations of fish were more heavily exploited than in previous centuries. By the time I came to walk around West Penwith, 80 per cent of the world's fish stocks were either fully exploited, over-exploited or had collapsed.

In the 1950s, British psychiatrist John Bowlby began publishing his ideas about grief as a universal instinct, an emotional response to separation that evolved through natural selection. Bowlby reasoned that the grieving mind swaddled images and memories of that which was absent; as the individual measured the outside world against these treasured and guarded memories and found the object of their desire still lacking, a

maelstrom of feelings arose. These sensations were generally gainful because they helped preserve an emotional and mental tie to someone or something of prior importance with which reunion remained possible. The occasions on which this preservation of memory was beneficial outweighed those when the permanence of the separation prevented reunion from ever taking place.

If grief stemmed from the expedience of keeping possession of one's own or others' former wisdom and intelligence, what repercussions arose from wilfully shearing off from the past?

Archaeological evidence shows that around sixty thousand years ago, Neanderthals and early modern humans began to bury those who passed away. Grave sites matured into landmarks that heightened the association between grief and the relationship to the past, gathering enough reverence across many human societies that they were preserved even by those of a different culture. Once the landscape bore traces of death rites, people wandering across tracts of country became aware of the mortality of both individuals and societies, the uncomfortable recognition of an earlier people's broken connection with the landscape. During the late Stone Age, around five or six thousand years ago, the people then occupying Britain began burying their dead ritualistically in the great chambered earth mounds of barrows and in giant stone tombs. These provoked deference among successive generations inhabiting the island, first the Celts and later the Anglo-Saxons.

The story of *Beowulf* returned to me as I read books on grief. Grendel, its monstrous anti-hero, reared up like a spectre from the nightmares of a conquering people, a phantom of the subconscious where the ghosts of those the Anglo-Saxons had felled lingered in poetic reckoning. Grendel was an uncanny form, somewhere between human and anarchic substance, plaguing the heaths and moors, one of the 'barrow-dwellers'. This giant inhabitant of watery realms stole into the settlement of Heorot in the dead of night, devouring unfortunate warriors who crossed his path. The hero Beowulf overcame Grendel, whose mortal wound betrayed him to his pursuer, the red flag of his blood across the water. The night after the victory, Grendel's mother avenged her son's death. The warriors followed her back to her unearthly lair, a pool flanked by sheer cliffs, overhung by a waterfall that concealed a cave. They entered the cave by diving into the water and swimming beneath the waterfall's tresses. Anxieties about the frailty of human societies barbed Beowulf's tale. Though both of the monsters succumbed to the blades of the men, the double deaths of Grendel and his mother ushered in the demise of the victors' civilization. As Grendel's grieving mother received her death blow, the heroes' victory gave way to a strange elegy, the 'Lament of the Last Survivor', which foretold of an unnamed man, the sole survivor of his people, who bore the burden of the tribe's history, with nobody left to inherit its riches. It was a vision of Beowulf's fate: from his ashes sprang the knowledge that his nation, unable to hold back the tide of its enemies, would die with him.

The Anglo-Saxons had migrated in several waves, searching for new homelands and resources, steadily destroying the cultures of the competing peoples whose lands they coveted. There is an argument that climatic changes along with the bubonic plague left some of the Celtic peoples susceptible to extinction. The research of palaeoecologist Dr Mike Baillie into the narrowed tree rings of Irish oaks suggested that matter from the Biela comet in the mid sixth century struck the Earth, causing 'a catastrophic environmental downturn'. In coming to Britain, the Anglo-Saxon tribes largely succeeded over their Celtic rivals, but in little over five hundred years, their own order would disintegrate through the toughness and expansionism of the Norman nobility of France. As they invaded, the Anglo-Saxon tribes came across the failures and excesses of their predecessors, the Romans, whose powers foundered partly as a result of their extravagant use of the natural environment.

Overcrowding, deforestation and soil erosion, along with the pasturing of herd animals, caused the exhaustion of soils and depletion of resources needed to feed Roman society and its subjects. Theophrastus said of Cyrene, 'There was an abundance of trees where now the city stands.' Strabo complained that the forests of Pisa were being felled for buildings and villas. It was increasingly difficult to find wood for shipbuilding, as superior local sources were soon exhausted. Pliny wrote that the deforestation of watersheds caused 'devastating torrents'. He also recognized that soil erosion damaged 'stones and earth, animals and plants'. Theophrastus even claimed that the clearing of the trees around Philippi dried up the waters and raised local temperatures. The *hylotomen* were professional tree-cutters, contracted to clear ground for farming and mining, and to provide stocks for ships and buildings, while government provided tax incentives for the timber trade. After several years of deforestation, harvests began to wane, the humus exhausted. Efforts were made to obtain supplies from further afield and conservation policies were instigated to salvage the last stands. Some places even embarked on strategies to encourage regrowth. The schemes were fatally optimistic and inadequate, instigated too late to be effective. As a consequence of their excesses, starvation and violent grievances spread across the empire, eventually bringing the downfall of Rome. We weigh heavily on the world, the Roman theologian Tertullian noted in the late second century; as our needs grow larger, so do our complaints that nature fails to sustain us.

Perhaps for this reason, the sentiment of nostalgia gripped the early English as they made their lives amid the remains of the Roman Empire. The ruins of Rome's former might must have intrigued those who came to inhabit the areas left by the retreat of Roman control. The Celts, who had the surest comprehension of Britain's natural history, the glitches of its weather, the tricks of its rivers and seas and soils, were often ill-fated adversaries of the Anglo-Saxons. As they seized their lands, the priceless knowledge of the Celts was sometimes forfeited. The language and poetry of the Anglo-Saxons communicated their uneasiness about pushing a people and culture towards extinction. *Wæl*, the Old English word for 'slaughter', spawned a disquieting number of words by annexing different endings. The sheer fertility of that root word communicated the brutality of the times as little other historical evidence could. *Wælsleaht* became the word for a battle; *wælspere*, a deadly spear; *wælgīfre* was a kind of slaughter-lust, while *wælgæst* stood for a murderous spirit – the phantoms that arose from the ashes or restless corpses that issued from unquiet graves at nightfall.

Yet there was little sign that this uneasiness prompted insight into the causes of decline. An ambiguous emotion, nostalgia was fatally contradicted by different, sometimes irreconcilable compulsions.

In his *History of the Wars* in Byzantium, written around the mid sixth century, the author Procopius noted that the only peoples living in Britain by this time were the Anglo-Saxons. While this was not in fact true, he recorded the seemingly widespread belief that the souls of the dead also populated Britain, ferried across the Channel in all their slight and shadow.

I spent the day alone at Priest's Cove. Above me, the skies were blue but for the tightrope of the horizon balancing a giant rhombus-shaped raincloud. The scuffing of the waves across the riveted stone of the cove was like a dare. The urge came on me to enter the water. I stripped to my under-wear, nakedness and the thrill of abandon turning my skin to goosebumps. I waded in up to my thighs and plunged, bowing my head beneath the small waves of my resolve. The water was achingly cold. Swimming out, I felt as though I might go on for ever, across the broad ocean – and beyond, over the edge of an outdated map, away into the midnight boundlessness.

I'd come across the name 'Priest's Cove' before, in Bottrell's *Traditions of West Cornwall*. Bottrell described a game, com-mon among the people of West Penwith, where a young man enacted the role of a dead lover. The man would linger outside the door of the room in which a woman sat, waiting. He would knock and enter. Once inside, he would threaten to carry off the woman unless she accomplished a sequence of tasks. It sprang from an ancient belief that the dead came to claim their former female companions, tamed through the centuries into light-hearted recreation. The game had a related ballad, 'The Grey Cock', in which a man who had drowned at sea returned to the home of his former love. 'The clays have changed me, I'm but the ghost . . .' he cries. This was the old tradition of revenant ballads sung to prick the conscience of the living. In these haunting melodies of death, the vanished escaped their graves, roaming the grounds of their former lives until dawn banished them. During my reading, I stumbled on something else too. These revenant ballads sprang up across Britain during the Norman Conquest, revived again by Romantic poets at the onset of industrialization. They were symptoms of disquietude, worried dreams of lost pathways to the past, expressive fears that something of great significance and value was irrevocably absent.

One of the early symptoms of nostalgia recorded by Hofer was the sufferer's belief that ghosts had visited and spoken to them. For such sufferers, the ghost arose from the permanence of death, its voice an echo rebounding against the hard reality of the world. This echo or phantom was the aftermath of extinction. It was all that remained for listeners to fathom, a message that they'd make of what they could. The idea recurred in the work of the Swiss psychiatrist Carl Jung. His exploration of dreams and the psychological equilibrium of the human mind during the cycles of waking and sleeping led him to formulate his theory of the 'imago', a trace that emanated from something's absence.

In the ancient Greek myths and stories, the hero often emerged symbolically from darkness into light on his nostalgic journey home, fantastical ideas that spread into notions of human psychology. In *On Sleep*, Aristotle described aspects of the human mind. *Eknoia* was an unconscious state, the mind idling somewhere between sleep and wakefulness. *Euthuo-neiria* were the symbol-laden eruptions of the sleeping mind – dreams, fears and yearnings channelled into myth and, rarely, visions of the future. *Eidolons* were the apparitions or phantoms, taken by the atomists to mean the dusts and vapours given off by external objects, penetrating the pores of the body and causing dreams and fantasies. *Mantikē* were those endowed with a paranormal faculty to make sense of dreams and perceive both past and present. The word *alloiōsis* stood for the special alteration of dreaming minds, the sea change of the imagination. Aristotle's peculiar and unique vocabulary sought to express his belief that while the original cause of a phenomenon might disappear, a powerful effect could remain, from which all kinds of imaginings might evolve. The most skilled interpreter of dreams, Aristotle claimed, was one who observed resemblances, someone who could reconstruct the original of a reflection from fragmentary or distorted images in troubled water.

Orpheus was the companion to the Argonautic sea lords and the inventor of the harp, instrument of singing wood. His name derived from *orbh*, 'to separate', and touched on *orphe*, Greek for 'darkness'. His myth acknowledged the ocean's dual role as life-giver and taker, the dream-darkness from which life emerged but into which we could disappear irrevocably. One of the Greek myths in Ovid's *Metamorphoses* tells how King Ceyx is punished for his arrogance by being drowned in a mighty storm. In the story, Alcyone, the king's wife, learns of her husband's death through the dream guises of Morpheus, son of Hypnos, the God of Sleep. Through the dewy dark, on noiseless wings, Morpheus flew into the town of Trachis. There he lay down his wings, and took the form of Ceyx, deathly pale and naked. As he stood beside the poor wife's bed, his beard and sodden hair dripped seawater. Leaning over her, he cried, 'Poor, poor Alcyone! Do you know me, your Ceyx? Am I changed in death?' The king's folly led to his drowning but the chance still remained for those left behind to search out the error and guard against its repetition. His wife Alcyone failed, alas, and threw herself into the waters in the grief of her loss. Death came to her as to all failed dreamers who could not divine order from the ocean's chaos, seeing only blackness and void.

The story resonated with nostalgic suggestion. If the past was of service to the present, why were its messages so rarely effective? If the succession of old methods of survival by stronger, more powerful efforts provoked sentimentality, why did a feeling of intoxicating regret rarely seem to lead people back or, if it did, why was it often inadequate?

On the one hand, the extinction of earlier traditions and industries in Cornwall were symptoms of the unsustainable effects of industrialization and a growing human population. But I began to wonder whether our nostalgic awareness of loss paradoxically activated, even increased, our urge to innovate. People are not only fascinated by losses but recoil powerfully from them; they try to evade imminent restrictions or to exploit them rapidly in an individualistic way. In recent years, fisheries scientist Callum Roberts argued from historical sources that around one thousand years ago European fishermen began plundering the seas more heavily than local freshwater sources. Most likely, the need for fish as a source of protein among growing populations began to outstrip catchable numbers in the safer hunting grounds of the rivers, lakes and streams. Fishermen devised ever more powerful tools for their catches, revolutions that, by the nature of maritime activities, spread rapidly around the world. According to Roberts, the first historical reference to the bottom trawler was a complaint from as long ago as 1376 to King Edward III, bemoaning the use of 'a net so close meshed that no fish be it ever so small enters therein can escape but must stay and be taken'. By the mid nineteenth century, mechanizations had strengthened the capabilities of the trawling fleets beyond the dreams of those under King Edward's reign. Fishermen still using earlier techniques to fish complained that the trawlers were wrecking the sea bottom and dangerously impoverishing fish stocks.

The innovations of the fishing industry were partly prompted by recognition that the environment was under threat, which led to further destructive measures. The same was true of whaling. The blue whale had entered my dreams again, a symbol of nature itself and its originality, perhaps provoked by musing on the watery giant Grendel. Did the whale's demise belong to the natural order of evolution, or was it a totem of human illogicality? While reading about fisheries, I found photographs of the bizarre, gigantic industrial ruins of commercial whaling in the subantarctic islands of South Georgia. These ruins, like those of the tin mines, were transfixing. The revolution from steamships through to the pelagic ships of the 1930s made possible the hunting of the blue whales of Antarctic waters, which were faster than the species of the northern latitudes. The enterprise nearly brought extinction to several species of whale and led to the sudden collapse of the industry. In retrospect, it seemed like madness. I wanted to understand the lunacy of pursuing profit despite all warnings to the contrary. I resolved to visit South Georgia to see for myself the corroding history of whaling that made it such a potent place.

The Second Peregrination

South Georgia,
Antarctica and
the Falkland Islands

4

Whales

I left West Penwith in the autumn of 2007. I was travelling at the invitation of the British Antarctic Survey, sharing in one of their research voyages that would stop by the island of South Georgia en route to Antarctica. My principal reason for visiting Antarctica was to study the ruins of the whaling industry. I was interested in what these industrial graveyards told of the violent changes between my grandmother's era and mine. After a week's training at the BAS headquarters in Cambridge, I flew from RAF Brize Norton to the Falkland Islands with a group of scientists, field assistants and medics. In Stanley, the main settlement in the Falkland Islands, I was to board an ice-strengthened research ship, the *James Clark Ross*, rigged to sail the Southern Ocean.

'I can see no difficulty,' Charles Darwin claimed, 'in a race of bears being rendered, by natural selection, more and more aquatic in their structure and habits, with larger and larger mouths, till a creature was produced as monstrous as a whale.' Closing a lecture in 1883, Darwin's contemporary William Flower told his audience to picture some primitive, marsh-haunting animals with broad, swimming tails and short limbs. These, he said, would gradually become more adapted to the water rather than the land, altering into dolphin-like creatures of lakes and rivers, and eventually finding their way to the sea.

Forty years earlier, amid the bluffs of the Ouachita river, a physician from Arkansas uncovered the bones of a strange, gigantic creature that he fancied a huge lizard. The bluffs dated from the Eocene epoch between some thirty to fifty-five million years in the past. Contemporaneously, the nearly complete skeleton of the same kind of beast was puzzled out of the muds of Alabama and scrutinized by British naturalist Richard Owen. He confirmed it as a primitive whale, the basilosaurus. Snake-like and with a blowhole close to the eyes and greatly foreshortened hind legs unable to bear its bulk, the basilosaurus lived a fully aquatic existence while its body still carried the physical signs of a life on land. In the later twentieth and early twenty-first centuries, scientists began digging out an amazing array of skeletons from the Himalayan regions of Pakistan, slowly honing our understanding of the adaptation of these various land ancestors of whales to their oceanic life. Such fossil remains as well as rudiments perceptible in the bodies of living whales confirmed their history as predatory land mammals.

At different stages of evolution, some of these beasts mated and bore their young on the earth but slipped back into the dark waters throughout much of the rest of the year. Others hunted both on land and in shallow pools, their organs and senses swithering between dust and flow. In 2001, while scouting in northern Pakistan, the research team of Dr Hans

Thewissen, a professor of anatomy, found the bones of animals now considered the earliest progenitors of the whale – wolfish creatures that glimpsed the world from the tops of their skulls so that they could otherwise remain submerged in water, beasts not a far cry from the bears that Darwin had visualized. Possibly, as Darwin speculated, they foraged further and further into the rivers and seas, gradually adapting to a solely aquatic life, or perhaps changes to the abundance of their prey or threats from other ferocious land mammals forced each stage of the whale deeper and deeper into the waters. Whatever the causes, uncountable summers transformed these earlier animals into the extraordinary whales of modern times.

The British Antarctic Survey's ship *James Clark Ross* was anchored in Stanley. I joined the ship in late October for its preliminary crossing of the Southern Ocean to Antarctica. As I walked across the floating platform to its gangplank, I noticed workers unpacking recent shipments of goods in the shadows of an adjacent hold. I darted over to them and was amused to find huge cellophane cubes packed with cans of Coke labelled 'Coca Cola Polar'. A terracotta-coloured shipping container housed several hundred orange kitbags, each tagged with the Union Jack and our names. I found mine among them and carried it on to the ship, remembering a summer day in the gear store of the BAS headquarters, trying on fleece underwear and thick, windproof jackets. I took the top bunk in a cabin shared with two other women. The yellow curtains that gave me a little privacy were William Morris-like designs of exoticized woodland weeds. Once we were at sea, they began to inspire odd, contorted dreams of home.

In recent years, a number of scientists involved with the BAS had received international praise for their research into the importance of the Southern Ocean for global climate, analysing eighteen of twenty-four climate models for the assessment report of the Intergovernmental Panel on Climate Change published in 2007. Much of the science explored dynamic processes that characterize transition zones such as oceanic fronts – areas where two bodies of water with considerably different characteristics meet – or the jagged boundary between the sea ice and open water that fringes the Antarctic continent. One well-known boundary is the Polar Front. Along its edges, warmer, subantarctic waters touch the cold polar waters of Antarctica. Superstition and a sudden chill mark the crossing of its mist-blurred line, the water temperature plummeting by around four degrees. We made our crossing of this boundary on 29 October 2007, a fluid feature that many scientists regard to be the movable threshold of Antarctica. The following day the American Senate debated the possible ratification of the 'Laws of the Sea', described as 'possibly the most significant legal instrument of this century' by the United Nations secretary-general in 1982. Those nations signed up to it were party to common rules on navigation, fishing and the economic development of the open seas. The pursuit of deep-sea mining interests had kept American governments from adopting the treaty in the past and they failed to endorse it on this occasion too.

Shortly after crossing the Polar Front, I witnessed my first iceberg, its colossal separation from air and ocean, its lunar defiance, intractable against the elements. One façade was rough and quite white but as we swung forwards I saw its slope of aquamarine, smooth as a cheek.

We arrived at South Georgia in early November. The small number of buildings that lined the shore housed one of the British Antarctic Survey research bases and the government of these remote, ice-struck islands. One of the first individuals in the world to sight them was London merchant Antoine de la Roché, when strong winds blew his ship off course in 1675. But James Cook, a sea captain from Whitby in Yorkshire, became the first to land on the islands of his own volition when his ship, the *Resolution*, anchored in January 1775, flagging the islands as British. Shortly afterwards, they were named after King George III. Within a decade of Cook's account of the large number of fur and elephant seals encountered there, sealers from Britain and North America began visiting South Georgia, shipping thousands of seal skins to home markets and trading skins and oils with China.

I hauled on a pair of snow boots, some padded salopettes and a bright orange BAS jacket, and clambered from the *James Clark Ross* on to the jetty at King Edward Point harbour. The ship towered behind me, its shadow casting a giant trapezium across the snow. The name belonged in the past to a vessel of a quite different character, a Norwegian whaling factory

ship that hunted from the waters beyond these shores. On Christmas Eve 1923, that *James Clark Ross*, under the command of Captain Carl Larsen, was hunting for blue whales in the Ross Sea. Over two hundred whales were slaughtered that season, yielding seventeen thousand barrels of oil. This was the first time the whalers succeeded in flensing their entire catch alongside the ship, enabling them to stay at sea. Until then, whalers nearly always returned to shore to process their catch. All of the commercial whaling countries experimented with different methods to escape this inconvenience. Sometimes a wooden cutting platform was lowered on to the whale's carcass, on which the whalers balanced precariously, slicing rings of blubber as if carefully peeling an apple. A year later, engineers succeeded in fitting rear slipways of staggering length on to the factory ship the *Lancer*, enabling the vessel to process whales entirely at sea. In less than ten years, the blue whales captured were noticeably smaller than those taken in previous decades, suggesting that nearly half of the catch was sexually immature and that the population was in danger of disappearance. The large baleen whales reached sexual maturity at a leisurely pace, their pregnancies stretching on for around a year. The population wasn't adapted to replace itself under such threats.

Across the sound from King Edward Point was Grytviken, the ruins of the earliest shore whaling station in the Southern Ocean, established by Captain Larsen. Grytviken, or 'pot cove', was so named by the Swedish Antarctic Expedition in 1902 after the large number of cast-iron pots left on the beach, once used to boil up elephant-seal blubber. I made my way towards the ruins, accompanied by a couple of scientists and the ship's doctor. It was a dizzying morning. Sunlight embroidered every surface, such that I had to squint to see the snow-covered path ahead. Before leaving, one of the field assistants had urged me to don my sunglasses. In weather like this, he told me, the snow could blind a person not wearing them in a few hours, erasing all senses to its defiant blankness. I lumbered prudently along the snow-lipped edges of the

brash-ice, whose startled white waves hid the immense full-
ness of the elephant seals. In the distance, I could see a complex of giant rusted structures against the snow. Beyond these
ruins was the spire of a white and red church, the source of
a stream that wound down to the shoreline, where an old
whaler and harpoon were stuck in the sludge. The belch-
ing and huffing din of the seals lessened and, at first, all was
quiet. Then I took a few steps further and everywhere there
was the sound of dripping water, oddly sanctifying cadences
against a backdrop of furnace and factory.

The first thing that I encountered was a wooden building
that housed a small museum. Steps led up to its olive-green
entrance, where a cream plaque read OPEN. Inside I slowly meandered through its narration of Grytviken's whaling
past. The previous evening I had read the diaries of Willem Van der Does, who worked aboard the first *James Clark
Ross*. His memories resonated in my mind, images of the ship
dragging seven whale cadavers through the thronging snow
squalls. Under the fazing tilt of the waves, the blubber and
meat cutters crawled over the mouldering bulks, propping
themselves on the carcasses, hacking and slicing. They carved
off the whales' lips to reach the baleen, and chopped the great
tongue away with an axe. They found rudiments of rear limbs
invisible from the outside. 'The whales are creatures that are
doomed to disappear from the earth,' remarked Van der Does
in 1934, 'despite all legal measures to prevent it.'

Humans have occasionally captured or made use of stranded whales since prehistoric times. The first organized whale

fishery was in the Bay of Biscay, established by Basque hunters in the Middle Ages. These whalers pursued the slower baleen whales, bowhead and grey whales, flinging harpoons at them multiple times in order to kill them. After exhausting whale populations in northern latitudes during the seventeenth and eighteenth centuries, European and North American whalers expanded southwards, innovating new technologies to do so. One nineteenth-century company listed the needs of the whaling ships in arresting detail: food, from flour, vinegar, preserved meats and cheeses to souchong tea and chocolate. A complete set of crockery. A complete set of tinware, from pans to a ladle and oil skimmer. A tin-coated blubber-room lamp and cook's lantern. A huge range of tools for making ship repairs, anvils, axes, adzes, jack planes. General hardware, from coffee mills to handcuffs. And, of course, the specialized tools: blubber forks, blubber mincing knives, rigging leather, friction rollers, pikes, toggles, harpoons, and fluke chain rings.

The traditional whaling society of Norway pioneered many of the new devices that allowed the whale hunt to begin in the Southern Ocean around South Georgia. First came faster whaling vessels powered by steam, then the explosive grenade-harpoon, the shore stations, and finally the floating factory ships. At the heart of the onshore stations was the flensing platform, on which the dead bodies of the whales were winched for butchering. Blubber was sliced away and boiled up in tripots to produce oil. The skeletons of the beasts were mechanically hoisted into huge bone lofts and rendered into more oil in bone cookeries. Oil was stored in gigantic metal tanks. And any remaining parts of the beasts were rendered down for bone meal. Initially, the whalers dumped the bones and the unwanted meat in the bay in front of the station but soon the British enacted laws requiring full utilization of the carcass. This was the absolute industrialization of the hunting of the whale, and the mechanized mass appropriation to human needs of a complex living creature formed over millions of years.

After leaving the museum, I wandered around the ruins, trying to guess the purpose of each rusty structure. I struggled to imagine what it must have been like when still active, the smell of butchered and burning meat, the clinking sounds of winched carcasses, the daily chatter of the workers. A young British whaler called Davies encountered the *James Clark Ross* at sea near South Georgia in 1949. In his diary, he described the ship's siren rising higher and higher until it made the sailors' heads ache and their ears sing. This noise, he said, mingled with the dense fog to give the whole scene an eerie quality. The sights were so surreal that the trip seemed 'all a long dream' to him. Davies commented that it didn't take long for the realities of whaling to stamp out any sense of romance and adventure. Once this disenchantment hit the whalers, he admitted, they saw only a marine slaughterhouse, with uncertain prospects of a good bonus for such tough and bloody work. I crept through the iron shadows of a large building, its corrugated walls stripped to the sturdy bones of its construction. Water dripped from its metal joists, an uncanny timekeeper. In one of the far rooms, there was the vague, blanched outline of a human head and shoulders chalked into the rust.

Beyond this large building, freakish instruments and workings cluttered the landscape. There was an overgrown toothed object like a saw, curved and beautiful as a Norse carving. I could see chains under fragile bonnets of snow with links bigger than my hands. In the distance, the disintegrating platforms of mortuary metal were apparent, proportioned to the massive bodies of whales. A short walk away from the strandline, I stumbled across a large tarnished tank on which the words BLUBBER COOKERY were painted in faded white lettering. At the front was a bolted, square-shaped opening. I immediately thought of the pictures in my childhood copy of *Hansel and Gretel*, the entrance into the oven where the siblings shoved the witch who held them captive to satisfy her monstrous appetite. But perhaps most startling of all was one of several oil vats, a round structure of astounding proportions, bulging through time and disuse. Sunlight nudged into its snowy circular border, and I too nudged forward, amazed, appalled, snow falling soundlessly around me.

I was jolted from my thoughts by the surreal vision of a huge modern cruise ship cambering into the bay. Grytviken whaling station and museum had become a tourist attraction for those wealthy enough to experience Antarctica purely for pleasure. As several hundred passengers disembarked to begin their tour, they broke my solitude, and I felt ashamed of my irritation. After all, I had no greater right to be here, so far from home. As the crowds ascended the stairs to the museum, I slipped away to the church and pondered what any of us would take from gazing on all this ruination.

It was quiet inside the church, a simple wooden construction with a pulpit and several plain rows of pews. It represented a different time, a different set of values. It had been sent from Strommen in Norway at the request of Captain Larsen and the Norwegian Seamen's Mission to care for the souls of the whalers. Erected in 1913, it had little success in inspiring their faith. The men worked too hard to be harangued into godliness. Besides, their work was needed by those back home; it was a means of obtaining a decent wage and provided first the fuel for lighting and then additives for preserved foods like margarine. At the back of the church was the old whalers' library, a time capsule of once popular but now long forgotten books. Here the men sat, the blood and stench washed from their hands, reading books like *Død mann forteller*, the 1950 Norwegian translation of John Masefield's novel *Dead Ned*, absorbed in a faraway, fantastical world.

For a short few years in the early 1920s, whaling was so profitable that companies paid wages largely in dividends, which stimulated the rapid expansion of the industry. Paradoxically, at the same time, smaller catches of whales prompted investigations into their populations in the Southern Ocean. Research to further understanding of whales and their habitat began in 1925, funded by whale-oil taxes. Surely the industry must have had a sense of threat to the source of its wealth. In 1932, the number of ships and their greater efficiency flooded the market to the point of making it uneconomic for several seasons. This had a complex effect: the whaling fleet was trimmed back, while the whalers were goaded into improving the efficiency of the remaining ships.

The Times newspaper of October 1935 reported the dispatch of the *William Scoresby* to obtain information on the migration of whales. The work of this specially equipped ship combined with that of the *Discovery II*, also engaged in acquiring information, which they hoped would 'lead to measures being taken to prevent the depletion of the stock of whales in the South beyond a point at which whaling will become uneconomic'. Wartime left Europe with a shortage of cooking fats and oils, which released investment to refit the whaling fleet that had been temporarily usurped by the war effort. As whalers resumed their hunting around South Georgia, global recognition of the decline and endangerment of many of the world's great whales began. The International Whaling Commission came into being in 1946 to impose regulations and quotas internationally, but failed disastrously as governments capitulated in the face of lobbying from whaling companies, allowing unsustainable whaling to continue. By the time the British Antarctic Survey began its scientific work at King Edward Point in 1969, whaling was finished, the stations languishing and idle. The natural world couldn't withstand the formidable breakthroughs of modern industrial methods. After the closure of the whaling stations, visitors stole artefacts from the dead industry, steadily picking through the rubble like birds at carrion. In 1975, Grytviken was designated an Area of Special Tourist Interest and the number of visitors has increased ever since.

The simplest narrative for this place was that greed led to the demise of the industry, the cavalier pursuit of economic gain while the going was good. It was tempting to consign these mistakes to the past, severing them from the present as if they had no further relevance. But the decline of whaling also had its origins in strategies of survival that ignored the limits of nature. Similar behaviour still weakens and devours the natural world, masked by the increasing convolution of our economic and political systems. Surrounded by the ruins of whaling, I wanted to reinstate its history as part of the saga of progress.

★

I could not keep my mind from the two world wars while I wandered around the ruins of Grytviken. The industrial wreckage of commercial whaling made me ponder the growing destructive capabilities of our species and how the wars had affected both the mindset of several generations and the tools at their disposal. Trudging across the snow, my shadow dancing more agilely beside me, the haunting melody of Elgar's Cello Concerto played in my mind. In March 1918, after an operation to remove his tonsils, the British composer woke up and asked for a pencil and paper and wrote down the opening theme of his famous Cello Concerto in E Minor. He was one of many artists whose work was conceived in the aftermath of the First World War, in the knowledge that society had altered irrevocably. Fifteen years later, as he was dying, Elgar hummed the concerto's formidable opening theme to a friend and said that if anyone heard this tune being whistled on the Malvern Hills, they would know who it was. In the years after the Second World War, J. B. Priestley wrote his play *The Linden Tree*. The principal character, Robert Linden, is a history professor harried into retirement, and his daughter a cellist. In the second act, the strains of Elgar's Cello Concerto drift to the audience while she practises offstage. 'It is a kind of sad farewell. An elderly man remembers his world before the war of 1914, some of it years and years before perhaps . . . And he goes, too, where all the old green sunny days and the twinkling nights went – gone, gone.'

In my early twenties, I went for tea with one of the few surviving soldiers to have fought in the First World War, the first war to utilize new industrial technologies with unprecedented consequences. His name was Harry Patch, 105 years of age at the time, once assigned to the Duke of Cornwall's Light Infantry, for which he fought in the disastrous Battle of Passchendaele in 1917. He greeted me in uniform at the doors of the nursing home where he was living in the Somerset city of Wells. Called up at the age of eighteen, he received only a little training in the use of a Lewis machine gun and a Webley 45 revolver, and was on the front line by his nineteenth birthday, earning sixpence a day.

'We moved forward to Pilkem, the trenches north of Ypres,' he told me in his quivering voice. 'At Pilkem, we were more at the top, and so we crawled. If you stood up, you were shot. I came across a Cornishman. A bullet wound is clean, a shrapnel wound tears you to pieces. And he was ripped open from his shoulder to his waist: shrapnel. Pool of blood at the side of him. And, as we got to him, he says, "Shoot me." And before either Number One or I could pull a revolver, he died. Thirty seconds. I was with him when he died. I would have liked to have got his identity listed, but Number One had moved on; I was a spare part, I had to go with him.'

In August 1914, Ypres was a picturesque medieval town, razed to rubble in a few short years, the bodies of more than two million soldiers palled by the clays of the outlying lands. In the years after the First World War, Sigmund Freud began studying the men who survived the trenches. They had a compulsion to repeat themselves, to restate again and again the horrors of their experiences, almost as if they had learned the lines by rote.

Instead of heeding warnings about the growing destructive powers of our societies, people devised all kinds of mental and political contortions to keep clear of any restrictions and continue with expansion and personal quest. In his 1909 manifesto for Futurism, Filippo Marinetti declared that society should rid itself of the ill-omened debris of romanticism. 'Do you want to waste the best part of your strength in a useless admiration of the past, from which you will emerge exhausted, diminished, trampled on? . . . We declare that a new beauty has enriched the splendour of the world: the beauty of speed.' In 1918, Marinetti founded the Partito Politico Futurista, a fledgling party soon assimilated into Mussolini's Fascist movement. Although Marinetti later condemned the Fascist and Nazi admiration for the perceived glories of their national pasts, his ideology none the less shared Fascism's opportunism, advocating oppression and violence in the pursuit of personal or national development, with the faint abuse of Darwinism amid the ferment. Beauty existed only in struggle, he claimed; there were no masterpieces except those of an aggressive character.

Experiments in industry gave Europe and North America a sense of freedom from the strictures of direct survival from nature, an achievement that cultural debate over the centuries proclaimed not only as the right path for human perfection but as the natural right of mankind. In actuality, while these technological advances contributed to a general rise in living standards, they magnified traits in human nature that resulted in destructive acts. The technology of the internal combustion engine, developed through a string of separate inventors modifying and adapting ideas, revolutionized the First World War. With its invention, industrialization after the war led to a staggering growth in the automobile, metal and oil industries – and so, too, the expansion of the means of trade, transportation and communications around the world. Between the wars, the demands of military operations fuelled the development of jet engines. British RAF engineer Frank Whittle and the German engineer Hans von Ohain simultaneously designed early jet-powered aircraft, the Gloster and the Heinkel, both of which made virgin flights in the lead-up to the Second World War. By the time I was a little girl, I took the possibility of intercontinental flight almost for granted, as if the world had always been thus. But this innovation, as with so many that structured my reality, had sprung recently out of the conflicts of the twentieth century.

Fossil fuels and mineral deposits form over millions of years of the Earth's spectacular and sluggardly phases. Industrial might extracted these resources from the ground and oceans in a fraction of this time. This was progress inextricably linked to the natural world but which modified people's lives in such a way as to convince them of their final liberation from it.

The major powers involved in the Second World War needed to hasten and strengthen their use of the Earth's natural resources, from the Japanese seizure of Chinese territory rich in iron ore and coal to the Nazi efforts to command Russia's oil fields. It was no coincidence that the greatest death tolls occurred among those nations whose mineral and fuel resources were most fiercely coveted by the Axis.

Even before the onset of the Second World War, the pursuit of finite natural resources spurred on aggression. In 1928, Fritz Thyssen, heir to 'King' August Thyssen's industrial empire of steel, coal and iron in the Ruhr Valley, mulled over the prospects of the Nazi Party. During the First World War, his father's company, Thyssen & Co, had produced a million tons of steel and iron for the German war effort. The Ruhr Valley, flush with the nation's natural mineral resources, was to pay the debts of failed conquest. When, in 1923, Germany defaulted on its reparation payment, the French and Belgian governments sent their labour into the mines. It was an occupation that crippled the German economy, outraging men like 'King' Thyssen and Albert Voegler, chairman of the German Luxembourg Mining Company during the First World War and, by 1926, the head of Vereinigte Stahlwerke, the largest steelworks in Germany. His profits flowed into the coffers of the Nazi Party. As a member of the Association of German Industrialists, he helped to raise three million marks for some of the most atrocious discharges of industrial technology of the twentieth century. Meanwhile, Thyssen's wealth transformed a stone house in Munich into the Brown House, the Nazi headquarters. Along the black soils of the Polish borderland of Silesia, his Consolidated Silesian Steel Company exploited the slave labour of the concentration camps, an appalling act of profiteering made possible by the brutal incorporation of Silesia into the Reich and the seizure of all regionally owned factories.

Eventually, of course, Hitler's grip failed, and the once impressive stone structure of the Brown House was destroyed by the bombs of the allies in the autumn of 1945 – but not before the war had caused the death of over fifty million people, as well as the emergence of ever more powerful inventions. The authorities cleared the rubble of the Brown House in 1947. It remained an empty lot for over fifty years, until forward-thinking ministers selected it as the site of the Documentation Centre for the History of National Socialism.

The horrors experienced by those who lived through these terminal applications of industrialization were unimaginable to me, despite my best efforts to comprehend the past of my grandparents and their generation. Contemplating these fatal times, people scratched their fleeting reality on to the landscape, communicating with the unknown of the future, to prevent, in some way, the escalation of their terrors. The strange 'I am' chiselled into something granted more permanence, like the famous chronicling of the Black Death scratched into the walls of a Hertfordshire church: *1350 was pitiless, wild, violent, only the dregs of people live to tell the tale.* Or the names scraped on the walls of POW and concentration camps during the Second World War, such as that of the US Grenadier Ralph L. Adkins, who carved his name with a belt buckle. In this instance, the Japanese had captured a small group of American soldiers after the sinking of

USS210 *Grenadier* in the Straits of Malacca. The POW camp became Form 2's classroom in the Light Street Convent, the names of the marines preserved behind a Perspex panel. I had seen photographs of this modern classroom, in which children born in the 1990s sat in attitudes of nonchalance and indifference.

Official artists gave expression to nostalgia for this devastated world. John Piper painted Seaton Delaval Hall, an eighteenth-century castle in Northumberland ravaged by a fire early in the nineteenth century. He visited the castle in 1941, his mind fevered by visions of the bomb damage in London. His brush captured the stark black outline of once undamaged magnificence, while its ruination was conveyed by a chaos of blue, pale pink, scarlet, hazel, grey. A wartime painting by John Armstrong depicted the damage to a church in the Essex village of Coggeshall near the painter's home. An official war artist at the time, Armstrong had already used the image of ruined structures during the 1930s, confronting the violence in Europe in the wake of the First World War as well as the effects of the Depression years. The painting of Coggeshall church portrayed the fuselage of the building intact beneath the bomb damage, its bare bones a fragile shelter of lost purpose. All of these individuals, whether scrawling on a prison wall or the official artists of an era, acted on their native compulsion to leave behind cautionary information for future generations.

In Germany the Nazis made use of the nostalgic sensibility to manipulate the perception of their society by future generations. In 1934, the Nazi Party's architect and future minister for armaments, Albert Speer, handed to Hitler his *Theory of Ruin Value*, in which he recommended the construction of buildings in anticipation of their slow and aesthetic ruination, so that Nazism might leave behind grand ruins like those of the Roman Empire. During the 1960s, in exile from Germany, Speer retyped his memoirs, contemplating the early years of National Socialism. 'Hitler liked to say that the purpose of building was to transmit his time and its spirit to posterity. Ultimately, all that remained to remind men of the great epochs of history was their monumental architecture ... Mussolini could point to the building of the Roman Empire as symbolizing the historic spirit of Rome ... Our architectural works should also speak to the conscience of a future Germany centuries from now.'

At the surrender of the Reich in 1945, the country lay in stupefying ruination – bridges, railways, factories, towns and cities bombed to rubble. There was a photograph of Dresden that mesmerized me as a teenager when I came across it in a school history book. In the picture, the empty buildings looked like sandcastles silting to nothing under the tide. It was incomprehensible to me that my countrymen could have so profoundly crushed this city; even more unbelievable that Germany salvaged so much of its infrastructure in such a short span of years. The rebuilding was not comprehensive; some ruins remained right through the years of a divided Germany. Only after 1989 did the unified government decide to repair these to encourage a forward-looking nation at peace with the mistakes of the past. A woman who posted her experiences on the Internet in response to this last phase of reconstruction expressed what she felt wandering through the ruins of Dresden during the years of Communism: 'It was an eerie feeling to walk past the charred rubble on a daily basis – a steady reminder of what war can do . . . To me the piles of rubble were sacred memorials to the thousands of lives lost during that night.'

After the Second World War, the German author W. G. Sebald wrote of a possible reaction against the persuasiveness of rubble, one of repugnance and avoidance. What he called *Trümmerliteratur*, 'literature of the rubble', sprang up from the ruins of post-war Germany. He saw the genre as symptomatic of a culture too afraid of retrospect, a people haunted by the image of Orpheus, the man with too little fondness for the future. Instead of imagining the face of Eurydice, who he is rescuing, he glances back and she founders in the underworld. As Sebald saw it, throughout the 1950s, many European writers responded to the devastation of the war as their political leaders did, by progressive action, reconstruction atop the ruins, or by aestheticizing the horrors in a way that deadened their effect – measures to avoid, at all costs, the ramifications of looking back. From his perspective, this inability to confront horrors forced people to avoid such subjects altogether, or to invent histories that enabled readers to live destruction by proxy, an aesthetic of horror that Sebald considered abhorrent. He singled out a scene from the work of Hans Erich Nossack, whose novel *Interview with the*

Dead condensed these ideas into a fablistic vision of survivors gathered around a fire: 'Then one man spoke in his dream. No one understood what he was saying. But they were all uneasy. They rose, they left the fire, they listened fearfully to the cold dark around them. They kicked the dreaming man, and he woke. "I have been dreaming. I must tell you what I dreamed. I was back with what lies behind us." And he sang a song. The fire burned low. The women began to weep. "I confess, we were human beings!" Then the men said to each other, "If it was as he dreamed we would freeze to death. Let us kill him!" and they killed him. Then the fire burned hot again, and everyone was content.'

The compulsion to conceal the horrors of the twentieth century and to revive each country's growth defined the latter part of the era, the same period in which destruction of the natural habitat increased to hitherto unseen proportions. The hefty, mechanical weaponry of the war scarred or destroyed thousands of villages, towns and cities, roads, railways and factories, along with the soft bodies and faces of millions of men and women. But the grief-stricken societies of the late forties and early fifties witnessed rebuilding on a massive scale, restoring towns and cities remarkably quickly, often in new, modern styles, quite unallied to the past. These societies believed they were acting for the good of people in burying the past, but in so doing set in motion a phase of development that picked up pace and devastated nature.

Those who lived through the wars never forgot them. Whether they felt able to speak of them or not, memories from those years haunted them. My grandfather, who worked in bomb disposal in Burma, never spoke of his experiences. But for my grandmother, who was in the WAAF, the war breathed daily through her mind, as did the breezes of her rural childhood. She returned to both obsessively.

Throughout the twentieth century, natural losses and extinctions were a known and largely accepted reality, thanks to the persuasive works of Cuvier, Lyell, Darwin and others. Knowledge of extinctions for ever altered the perception of natural history. Before industrialization, people expected to inherit their parents' world; after its onset, people came to anticipate a different future from that of their forebears, an ideology that the concept of extinction aided. One of the defining changes brought about by the industrial technologies of the war years was that people took extinction of life forms and ways of life, of landscapes and cultures, as the norm rather than the exception. The industrial clout and fervour produced by the wars condensed slow evolutionary or environmental processes into sudden detonations and rapid, engineered events. Death and earthly transformation became an almost instantaneous reality. People were terrified of this modern, man-made energy for decades, and each new generation since has had to come to terms with its lethal potential.

Yet the bewitching ease of modern life diverted attention from the behaviour causing faster and faster destruction. Civilization removed people, as much as possible, from the threats of a life contingent on nature. From the frugality of wartime, societies evolved that were wholly estranged from the realities of where their food, water and fuel came from and what damage had been done in the extraction and transportation of these resources. The information from the natural world that might have caused people to reconsider the moral obligations of their new lives was simply invisible, obscured by distraction or distance.

As I walked back from the church at Grytviken, the sun had melted the snow around shapes of promised warmth where buckles and barbs and other unfathomable metal objects lay buried. These possessions unhinged from purpose brought to the scene the strange contrariness of a ghost town, the imposing sense of abandonment. The whole landscape reminded me of a picture taken by the American photographer Walker Evans during the Great Depression. Evans hoped that the mechanical eye of the camera would compensate for America's blindness to the difficulties and poverty of its people, inspiring social change. But it was not the images of farm workers and labourers that had struck me as I flicked through a book of his work shortly before leaving for Antarctica. The photograph that remained in my mind was a close-up of a piece of obsolete machinery, entitled 'Tin Relic', a new kind of beauty, almost like that of a flower, imposed on the skewed metal by Evans' eye.

Treading the line of shadow from one of the structures that obscured the sun, I stood before the whalers' barracks. Here the men emptied their heads of the day's slog. Unlike the manager's villa, which the authorities converted into the museum, the whalers' space was partitioned into crowded corners. Those of a slightly higher rank might gain a small separate bedroom and a dining room shared with fewer men, but the majority of workers – and certainly, those engaged in the longest hours of manual labour – slept beside one another in narrow beds and were crammed together at a workaday table for their hurried meals. While the library, the cinema and the football field offered some light relief, it struck me just how confined were the physical and psychical spaces of most of these men. When the whalers collapsed into their cramped beds at the end of the day, the smell of the oil from the processing plants penetrated their sleep, oil that had mystifying, dreamlike properties, infusing the lives of people thousands of miles away; it was there in the dynamite that miners used to blast into rock, in the explosives erupting the battlefields of Ypres, and in more prosaic products like the soap with which men and women cleaned their hands over kitchen sinks.

The diary of Davies revealed the diverse feelings and impressions of those working at Grytviken. In one entry, he

posed beside the flensing deck in the shadow of machinery that altered the living reality of the whales with predictable and undeviating might. The blue whales, he said, were as sleek and streamlined as some unattainable ideal for an American motorcar. 'They looked prosaic enough lying there on the butcher's slab as it were – cold and dead. But when we started chopping up they were transformed . . . Steam driven saws – two of them on either side of the deck – make short work of any morsel set before them and the boiler mouths gape and swallow up everything til nothing remains of the whale but the bloody deck.' On his voyage, the blue-whale hunt began on 15 December 1948, and they killed thirty-one on their first day and thirty-two on the second. I found these figures astonishing. Although we spent nearly two months at sea, I did not see a single blue whale in the Southern Ocean. By contrast, in the new year of 1949, Davies and his fellow whalers regularly caught 'big fat juicy Blues, fat with oil and bonus'. In this reflection, he recognized 'the financial aspect gaining ascendency over the aesthetic and romantic!'. The whaling season ended in early March, when sixteen thousand 'units' of blue whales had been caught by the competing ships. The men worked for around twelve hours a day on deck, increasing their hours as the end of the season neared and they were pushed to reach their targets for the investors back home. Between the start of the twentieth century and the 1970s, when whaling finally ended, men killed around 1,500,000 whales in the Southern Ocean of the Antarctic region; roughly 300,000 of them were blues.

In a paper from the 1980s analysing the economic history of whaling, Colin Clark argued that in societies ordered on the basis of monetary gain, the exhaustion of a natural resource prone to extinction was a rational outcome. I found his ideas highly provocative. According to standard economic theories, he explained, the collapse of whaling stemmed from the absence of jurisdictional rights on such creatures; ownership of a killed whale was established by capturing it. Indeed, the floating factory ships were developed partly to evade any shore constraints imposed by the British governing South Georgia.

Given the potential profits from whaling and the finite nature of supplies, some degree of cooperation among competing fishermen should have materialized. And yet, said Clark, any such concords failed to prevent the near extinction of a number of whale populations – the grey and bowhead whales of the North Atlantic, and the blue and fin whales of the Southern Ocean. One of the cooperative efforts came after 1931, when whaling became uneconomic. For a few seasons, a gentleman's agreement set catches according to Blue Whale Units (BWU). The goliath among whales could measure as much as thirty metres in length. One BWU equalled a catch of two fin whales, six sei whales or thirteen minke whales. After this first informal quota for catches lapsed, the International Whaling Commission entered the picture in the hope of regulating the industry. It specified the figure of sixteen thousand BWU per season. In effect, this indiscriminate limit only motivated the whaling companies to compete heavily, boosting the effectiveness of their technologies. In five years, the same number of units were caught in half the number of days, little over two months.

Economic theory assumed that an equilibrium would eventually arise between the rate of exploitation and the natural production of the resource. The blue whales and other baleen whales bred and nursed their young in tropical waters, migrating to the Antarctic for the tiny crustaceans and blooms of phytoplankton that thrived at the ice edge. They feasted during the austral summer, then returned to their nurseries. The hope was that a sustainable harvest of these creatures would permit the survival of the whaling industry. Clark argued that firms engaged in whaling on the basis of anticipated future profit ultimately gained a greater return on their investment by exploiting whales to the point of extinction than by operating under a yearly quota, especially when a species was slow-growing. For the sake of argument, he wrote, suppose the stock consisted of 200,000 BWU and the sustainable harvest was ten thousand blue whales. If the entire whale stock was captured and sold, the invested return per annum would outstrip income from the annual yield. Clark proposed that perhaps property rights on whales, such as those conferred by the 'Laws of the Sea', might mitigate this outcome. For me, what linked his theories with other human-caused extinctions or near extinctions of natural resources or species was the unambiguous evidence that awareness of restrictions in the environment fostered greater exploitation.

Why, in an era in which our rate of exploitation was so

threatening, did the powerful nostalgic compulsion to re-consider and conserve fail to gain ascendance? Belief in the civilized superiority of humankind and of the intrinsic en-hancement of our species had driven the desire for libera-tion from natural constraints. As they weighed human nature against these technological inventions, industrialists began to dream of liberating industrial and economic activities from the constraints of human nature too.

In his diaries Willem Van der Does reflected on the poten-tial among the whalers to regret the slaughter of the animals. 'The death struggle of this enormous animal was dreadful to behold,' he said. 'His unbelievable dimensions notwithstand-ing, he was in fact completely defenceless. Involuntarily the men felt a deep sympathy for these gentle giants when they were finished off so mercilessly.' The instinctive compassion of the whalers described by Van der Does became an inef-ficient and restrictive aspect of the natural world.

From the outset of industrialization, societies measured people's capabilities against those of machinery. Physicians began to classify the symptoms of fatigue and to prescribe methods of enhancing the efficiency and yield of workers. In the nineteenth century, the French physician Etienne-Jules Marey used time–motion photography as if to capture on film the mechanical potential of human beings. Through his studies of blood flow, he blurred the line between the pump-ing heart and the orderliness of mechanics. His comparative anatomies of humanity and artificiality raised questions about how, in keeping with technological progress, people could

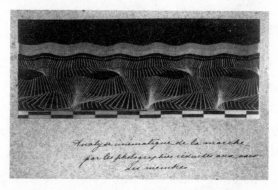

function to greater advantage. By the early twentieth century, the prevailing ambition among manufacturers was to organize labour for maximum efficiency, theories often modelled on the factories of Henry Ford and his vast assembly line of automobiles, muddling men further with mechanical possibility.

During the twentieth century, the productiveness of the world's labour force rose, seeming to bear out these earlier designs, whether through the hours of work, the flow of temporary workers, or other, invisible mechanisms and incentives. Still, the urge to fuse humans and technology persisted. As the theorist Norbert Wiener predicted in the 1950s, 'We have modified our environment so radically that we must now modify ourselves.' Today, the World Transhumanist Association purports to encourage the sensible use of technology to increase human potential. 'We support the development of and access to new technologies that enable everyone to enjoy better minds, better bodies and better lives. In other words, we want people to be better than well.' Like Nietzsche's *übermensch*, his more-than-man described in *Thus Spoke Zarathustra*, these will be humans who dare to burst through the bounds of their humanity.

Pursuing profit under increasingly restricted circumstances, the greater efficiency of machinery slowly edged people out of work; or else they were caught in systems beyond their control, where their patterns of working were organized for maximum efficiency and never for taking stock. Unlike machines, which operate with unswerving efficiency until they malfunction, the minds and actions of human workers are whimsical, impulsive and unsettled. Exhaustion and rigid, oppressive occupation dull the mind's innate capacity to weigh up the world and think again.

These days, the waters that surround South Georgia are a strictly guarded marine zone, and the provisions of the 'Laws of the Sea' control the fisheries operating in the region. Other areas have also adopted marine protection zones to reduce the likelihood of such disastrous over-fishing, although far greater measures are needed to slow the rate of exploitation. In the ancient past, human hunters did utilize a natural resource to the point of extinction, and presumably their unsentimental slaughter of creatures was akin to that of any other animal. Indigenous hunting societies still extant today have strict and respectful taboos to reduce the likelihood of harvesting beyond a sustainable level, but also societal norms that quell the compassionate sentiments of those who survive from nature. These small societies are distinct, though, from the prehistoric hunters, in that their cultures are often responsive to endangering populations of animals from which they derive their way of life. The whalers of industrialized countries operated under more bewildering conditions, compelled by natural impulses to exploit but acting from motives that were skewed by layers of history and development. The punishing character of the work crushed any germinal doubt or query, while the emphasis on economic investment and a free-for-all approach to the whales increased the likelihood that any inkling of diminishing supplies would only stimulate the exploitation.

5

Ice

After I left Grytviken, the Southern Ocean became impulsive and unpredictable. Sometimes the seas were rough and the ship's metal keened; at night the arrow tips of a snow squall rushed like a million migrating souls through the ship's beams as they searched for icebergs. At other times, the seas were astoundingly gentle, pacifying the giant circumference of the horizon as far as I could imagine. In such stillness, everything surrendered to inconceivable detail, the directionless, almost cellular agitation of the waters revealed briefly by a sunbeam. I lay in my small bunk on the *James Clark Ross*, listening to British folksinger Lal Waterson singing, 'Sleeping in my bed, strange thoughts running through my head . . .' In my hands was a book borrowed from the ship's library, Mann and Lazier's textbook *Dynamics of Marine Ecosystems*. The world's oceans had their own patterns of circulation and partition, the authors explained. Bodies of water quietened and enlivened, clashed and uncoupled. They hardened in conversation with the air, releasing salt into the depths, changing the nature of the deep waters below.

Occasionally, I pulled my earphones away and chatted to Claire, a penguin scientist from Yorkshire who was sharing my cabin. She was embarking on a four-month research trip, along with Ewan, a young Scottish scientist packed for a two-year stint studying seals. Others on the ship were on their way to investigate a variety of subjects from whale and krill populations to the frozen soil beneath the snow and ice. I was to stay on board the *James Clark Ross* for several months, through to our final destination, the Antarctic Peninsula, weighing up the ways in which the work of the scientists in Antarctica touched on the subject of extinction.

I was feeling excited about our next port of call, Bird Island, one of the small islands off the mainland of South Georgia, the breeding ground for several species of albatross. The growing endangerment of these large, spectral seabirds brought me round to a different sequence of ideas. The land-scapes of Cornwall and Grytviken were physical proof of some of the ways in which people played havoc with nature. By contrast, the possible demise of the albatross was more incidental. For this remote population of birds, there was no tradition of harvesting that might alert us to their endanger-ment. The practice in the Falkland Islands of gathering their eggs had been banned, so their salvation resided in our will-ingness to pursue knowledge of them as an end in itself.

While I was away, a friend of mine forwarded an email about the American photographer Chris Jordan, who took pictures of albatross chicks that had died and decomposed to reveal the plastic trash in their stomachs. Jordan called his sequence of images 'Intolerable Beauty: Portraits of American Mass Consumption', explaining that 'Exploring around our country's shipping ports and industrial yards, where the accumulated detritus of our consumption is exposed to view . . . I am appalled by these scenes, and yet also drawn into them with awe and fascination.' A debate about the photographs took place on the website of the *New York Review of Books.* One blogger's rhetorical question stuck in my mind: 'I wonder what future archaeologists are going to think when they find this site with obvious human trinkets, and some bird bones.'

Jordan photographed the dead chicks at the Midway Islands, an unincorporated part of the United States in the northern Pacific Ocean and the site of a notorious battle during the Second World War. The islands had an airport and a human presence. But people have not impinged on the albatrosses so directly at their nesting sites in South Georgia, where only small teams of scientists inhabit the research stations under strict guidelines to protect the birds. Despite the enthusiastic efforts of the scientists researching at Bird Island, all of the species of albatross breeding there – wandering, black-browed, light-mantled and grey-headed albatrosses – had been diminishing in numbers since the commencement of the studies on South Georgia, particularly as the long-line fisheries pushed further south into the Southern Ocean in the 1980s. The perils that faced the birds lay hundreds or thousands of miles out at sea. By the beginning of the twenty-first century, scientists found the same pattern of population decrease among all but two of the twenty-two species of albatross scattered around the world. As fishing fleets ventured into deeper, rougher seas, these birds encountered ships so commonly that they habitually followed them, swooping down to feed on baited hooks, snagging on them and drowning, leaving their chicks to starve. Those adults that survived sometimes returned to their chicks with mouthfuls of human debris, accidentally murdering their offspring. The authorities in South Georgia placed strict controls on the waters under their jurisdiction and this had greatly reduced albatross deaths around the islands. But even the most isolated albatrosses forage at sea hundreds or thousands of miles from their nests, and hence are threatened by our societies, regardless of their apparent distance from our reach.

The name 'albatross' derives from the Spanish and Portuguese word *alcatraz*, once used by sailors for a number of large seabirds like petrels and gannets. John Fryer described 'albetrosses', during his voyage to India in the seventeenth century as 'those feathered harbingers of the Cape'. Later William Dampier observed the 'algatross' and, in George Anson's 1740 circumnavigation of the globe, it was called the 'albitross'. English privateer George Shelvocke's *Voyage Round the World by Way of the Great South Sea* yielded an anecdote about the albatross that the poet Samuel Coleridge later immortalized in his 'Rime of the Ancient Mariner'. After heading southwards from the Le Mair straits, Shelvocke and the other sailors hadn't seen a single seabird but for 'a disconsolate black Albitross'. The bird seemed to hover about them as if lost. Finally, in a fit of melancholy, Halley, the second captain, shot the bird, declaring that it was a bad omen and that its death might bring good weather. In Coleridge's imagination, it had the opposite effect. The shooting of the albatross was 'a hellish thing', and the commanding association of the bird with the elements wrought havoc on the sailors' lives and fortunes.

> *Ah wretch! said they, the bird to slay,*
> *That made the breeze to blow!*

As we neared Bird Island, half a mile off the mainland of South Georgia, the winds grew stronger and the sky became a scaffold of clouds in greys of iron and steel. The *James Clark Ross* sought shelter in the bay at Elsehul, where the crew dropped the anchor, and the ship heeled and pitched on the waves. In the distance, we could see the raked outline of Bird Island, only its highest reaches soothed by snow. James Cook sighted the island on his second voyage in search of the fabled southern continent in 1775. Cook didn't think much of the tiny island or of South Georgia in its entirety. 'Wild rocks raised their lofty summits,' he commented, 'til they were lost in the clouds, and the valleys lay covered with everlasting snow.' Before the sun went down, I stood out on the deck watching black-browed and light-mantled sooty albatrosses carving the air, sweeping upwards, lassoing the sky. And white-chinned petrels as black as the night, harbingers of the oncoming darkness. There had been a silvery edginess to the day, as if the world was unnerved by the gathering weather. The ocean was an ominous, gunmetal grey, and a single wandering albatross drifted alongside the ship, its wings tipped to the sea as if to overhear the waves.

The following morning, the weather was clement enough to allow us to take the smaller boats ashore. A massive, contorted, bright blue iceberg challenged the entrance to the bay, and penguins and seals nudged up the waves, then quickly slapped down under the waters. Ahead of us, the beach was jostling with aggressive young male fur seals. Landing at a small wooden platform, we formed a nervous line, edging past the snarling and huffing mature bull seals towards the little research base at the head of the beach. The sealers who passed the island in the interwar years observed no more than a handful of fur seals, which were, by then, nearly extinct through hunting. Isolation and protection has since enabled their revival to such an extent that the precious stumps of tussock grass rising from the shoreline of the island are being eroded as the creatures push beyond the bounds of their overcrowded littoral territories.

Bird Island was the wildest landscape I had ever experienced. The prolific stench of animal life, dead and living, made me giddy. As I hiked up a frozen stream the racket of survival on the beach below gradually faded to nothing. A pair of black wings, like those of a fallen angel, lay in the middle of the snow-covered stream, its hinges of gristle still intact. As I ascended the hill, I seemed to stride back through aeons to the primeval past. On giant clumps of dead and living tussock grass were the wondrous aboriginals of the young wandering albatrosses. I stared in amazement at the old beardedness of their white plumage, the snare of their long, hooked beaks, the perceptiveness of their round, pitch-black eyes. Although the ancestors of these birds probably began to twist into shape over fifty million years ago, the first confirmed albatross fossils date from the Oligocene era, around thirty million years later. Between the Eocene and the Oligocene, the extinction of large numbers of plants and animals took place, especially marine creatures like archaeocetes, ancient whales. But the diversity of life gradually recovered through the many millennia of the Oligocene and new forms appeared, some of which eventually evolved into the whales and albatrosses alive today.

During this time, the land masses of Antarctica and Australia slowly wrenched apart, altering the world's oceans. The isolating flow of the great Antarctic Circumpolar Current swirled into force around thirty million years ago, establishing the polar fronts and the icy quarantine of the great, heart-shaped southern continent. Ever since, the size and extent of the Antarctic and Arctic ice caps have changed the traits of the oceans, bringing about adaptations both physical and territorial to seabirds like albatrosses. I found it a curiously exciting thought that albatrosses once took flight from shores closer to home. In the final decade of the nineteenth century, the British geologist Richard Lydekker scraped some bones from Suffolk crags around two million years old that he named *Diomedea anglica*, the English albatross, an extinct bird somewhere between the short-tailed and great albatross species. The bones were evidence that some species of albatross once bred in the North Atlantic but that the transformations of the Earth forced them to retreat steadily southward. Just imagining this prior reality set the compass of my mind spinning.

Bird Island was a trackless and tempestuous landscape, a wild land where natural forces reigned. As I stood on the mull of the hill, the fragile fortifications of the icebergs holding their line against the stormy seas below, my mind turned even further back, to riotous landscapes of prehistoric forests, jungled and unbroken, to times when nature alone transfigured the Earth, not the technologies of humanity. Over 400 million years ago, the early clutch of exotic plants began their lustrous capture of the landscape. The first, monstrous forests emerged from giant horsetails, ferns taller than several men standing on one another's shoulders, and the simple, olive-green club mosses. After this grew the beautiful cycads and conifers, their nuts and kernels trumpeting their fertility, and giant ferns that sheltered little revolutionary blisters of pollen. These presaged the flowering beauty of our world, such as the exquisite, magnolia-like *Archaeanthus linnenbergeri*, ninety-eight million years old, with its six generous curls of petals and the dissected hearts of its leaves. And the buzz and dart of insects and birds, their lives strung on the scents and dusts and fruits of these early flowers. No such wildernesses as these exist anywhere in the world today, not even on Bird Island. Our influence is too pervasive. But the Earth's astounding potential is there, recorded with almost photographic precision in its rocks, soils and ice. From these, we know of times and landscapes absolutely outlandish to our imaginings.

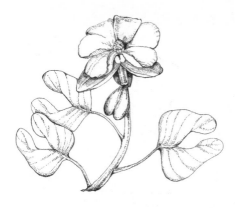

I wanted the albatrosses to survive here on these craggy subantarctic islands. But thinking about the English albatross was unsettling. In my mind, since my childhood, these regions represented the least tamed and mortal places on the Earth, yet the inhospitable conditions – the forbidding cold, the uncultivatable ice and the unstoppable winds – hadn't existed in earlier times. This austere landscape had its own history of alteration, long-ago ages when exotic plants and creatures thrived, then perished. Like the nineteenth-century geologists who mused over ancient, tropical forests etched into the rocks of their homelands, modern scientists studying in the Antarctic regions have unburied evidence of lush trees and foliage. From its rocks, we know that Antarctica was once part of the supercontinent of Gondwana, until the eternal process of sea-floor spread broke up the giant land mass, lodging Antarctica across the South Pole over sixty million years ago, before the onset of a long-drawn-out phase of cooling. Scientists drilling as deep as a thousand metres into the floor of the Southern Ocean have scrutinized the mud and rocks, glimpsing this summery landscape over fifty million years ago, utterly iceless and habitable, covered in forests and shrubs. The fearful, titanic ice sheets of Antarctica that I had imagined since childhood only grew as the world's levels of greenhouse gases declined around twenty million years later. The very leafiness and floweriness of the world seemed to have beckoned this ice.

Recently, geologists rooted out six forests from the ceilings of a series of coal mines near Kentucky. Dating from long before the birth of Antarctica, they were some of the first forests to grow on Earth. During the ages when these forests were evolving, the climate altered from one of large ice caps to a green world in which ice was only a memory. The carbon dioxide in the atmosphere surged and plunged throughout the eras, hindering and revitalizing the ice sheets at the poles. There is evidence to suggest that the processes of photosynthesis that took place before land plants evolved didn't draw enough carbon dioxide from the atmosphere. Carbon dioxide arose naturally from sea surfaces, volcanoes and the respiration of life, but over the millions of years that flowering plants flourished on the Earth, levels gradually lowered and the world's climate cooled. The shimmering surfaces of the ice sheets formed, slowly establishing the blizzarding climate of Antarctica. I asked myself which was its true nature: the epic ice that affected climates around the world, the flowering continent of earlier eras, or an Earth where Antarctica did not even exist?

Some albatrosses have been native to the frosted reaches of the Southern Ocean for tens of thousands of years. The same inventions that gave us insight into the demise of these birds and their wild province also allowed a vision of time and nature from which no easy conclusions could be drawn. These technologies gave a truly humbling breadth of perspective, an awareness of times when the world was utterly different, exposing the arbitrariness of maintaining the status quo.

I had been searching for critical moments of closeness to the natural world, junctures to which, perhaps, people might return to mitigate some of the damage to nature in the modern age. But my thinking was based on naive ideas about what was real and natural and wild. Were there really times when our relationship to the Earth was natural to a lesser or greater extent? Were there times, for that matter, when the world itself had been more or less 'natural'? I was seeking some essential difference between natural and man-made worlds and the point at which we became severed from nature, but the wildness of Bird Island and the natural processes that had formed the place only confounded my search. The primordial tropics of this terrain subverted any absolute idea of what was natural. Here, in the wildest place I had ever known, with only the slightest presence of people, the landscape spoke to me less of some undisturbed, remote stage of nature to which we might get back, than of the turbulent forces of change, the indefatigable action of time that made it impossible to fix on a point for return.

After visiting Bird Island, I thought more and more about degrees of greater and lesser wildness, and what significance they had for our future relationship with nature. John Muir, who emigrated with his family from Scotland to Wisconsin in the United States in the mid nineteenth century, once argued that conservation was 'not blind opposition to progress, but opposition to blind progress'. He became an inspirational figure in the rising conservation movement of his age, especially through his concern to preserve the great forests of his adopted homeland. He was a forefather of the Wilderness Act of 1964, which defined wilderness for the American public as 'in contrast with those areas where man and his own works dominate the landscape', and one of Britain's largest conservation organizations, the John Muir Trust, was named after him. His deliberations led to the common definition of wilderness as something in its original natural state, but meanwhile few places now remain unmarked by human presence.

Some definitions of wildness include remnants of past human activity like agriculture. Such archaeological remains undergo a natural change, by virtue of their redundancy and their weathering, until they return to the ground. But incorporating the sites of past human activity into the concept of wildness suggests that some means of survival are more natural than others. In recent years, a nostalgic interest in reviving small-scale agricultural practices has emerged, as if earlier methods of farming are truer to nature than those that are more mechanized. During the Second World War, American scientists laboured to develop a kind of wheat that would be resistant to disease and yield more per acre, to assuage the poverty and hunger in Mexico. Such developments were the makings of a new revolution, said William Gaud, the Director of the US Administration of International Development. Gaud called it the 'Green Revolution'. In the 1960s, as India and Pakistan struggled to feed their enormous populations, the use of engineered crops – along with pesticides, fertilizers and new irrigation methods – nearly doubled the crop in five years in both territories. These agricultural innovations largely averted starvation across these countries. They also took a considerable toll on the environment, reliant as they were on only a few varieties, along with chemicals, oil and vast amounts of water. Some people protested, such as Vandana Shiva, an Indian woman who spearheaded a conservation effort to oppose the use of 'engineered' crops. Her organization, Navdanya, advocated the protection and perpetuation of a wider variety of plant species. From her perspective, intensive agriculture and genetic engineering threatened not only her country's soil and plant diversity but also the native grip of her people. For Dr Shiva and others, agriculture that used only finite regional materials should be understood as wild.

But Dr Shiva and others engaged in more modest traditions of farming contributed only a small amount to the immense demands for food. In 2000, an Oxford University professor, Paul Collier, argued that the nostalgic yearning for the old ways of cultivating the land was the indulgence of wealthier nations. 'Europe can afford romanticism,' he argued, 'but the African poor cannot.' Africa and other countries likely to suffer the most from climatic shifts and natural disasters will need ever greater technological intervention to produce food unless populations decline markedly. Genetically modified crops designed to resist drought or flood might mitigate the risk of famine. In 2001 the European Union reported that they could find no risks to people or the environment in over fifteen years' worth of study into genetically modified crops, yet many countries voted to uphold outright bans or major restrictions. People were repelled by the idea of genetically modified food, as if something intrinsic might be perverted with unforeseeable consequences.

There was an associated anxiety about the threat to the wild originals of cultivated plants. Many fruit and nut species originated in countries like Kazakhstan, Uzbekistan and Kyrgyzstan, and in other mountainous regions of central Asia. A number of species became threatened with extinction due to people clearing forests, logging, building and using chemicals for agriculture.

By recognizing the difficulty of establishing a point of naturalness in a changing world, I had to define what I believed true nature to be. Within the magnificent proliferation of life and variation unforced by people, there exists a spectrum of what is natural. From one perspective, something can be natural through its integrity. Aristotle argued that natural things came into being under their own powers. The thing's intrinsic nature bridled its potential and activities. The wild originals of cultivars once altered only in accordance with nature. But after the onset of agriculture, people began to guard them against natural elements that influenced whether they flourished or not. In this way, people forced the plant's growing potential beyond its ordinary response to the environment. But was this an unnatural act? In *Nature*, John Stuart Mill challenged Aristotle's definition. From his perspective, changes to nature provoked by human technologies remained natural, for the nature of a thing comprised everything that affected it. Darwin acknowledged that animals influenced the form of plants. In *On the Origin of Species*, he pondered whether the rarity of a kind of bee might provoke a flower like the red clover to adapt to snare the bee more easily for pollination. 'I can understand,' he said, 'how a flower and a bee might slowly become, either simultaneously or one after the other, modified and adapted in the most perfect manner to each other, by the continued preservation of individuals presenting mutual and slightly favourable deviations of structure.' If the bee made changes in the flower or the flower induced changes in the bee, why should people interpret their own steering hand as any less of a wild force, even if they interfered with the genetic kernel of an entity?

Across large expanses of time, the natural world had no undisturbed, steady state; motiveless events altered life on Earth. Smaller stretches of time allowed other forces to predominate. Natural change took place through interactions among organisms and environments, from competitive encounters and oppositions of living things to the effects of a dominant species on its environment and those around it.

Early conservationists like John Muir interpreted wildness in nature as benign or even positive. These assumptions infused debates about the best way to moderate the damage people caused to nature. In terms of our species, there were fewer affirmative ideas about a return to a natural state. I began to explore the implications of perceiving our own nature in a negative light, for muddled with this were doubts about a life close to nature. Metaphorically, to be 'in the wilderness' was to be in a wretched state. Wildness, among people, implied a dissolute character and was associated with ideas of savage or barbarous tribes. In a disparagement of the Irish in the twelfth century, the clergyman Gerald of Wales linked the unrestrained growth of woods in Ireland with the failure of the Irish to progress fully from their former primitiveness. 'This nation despises agriculture,' he said, 'and is averse to civil laws.' The Irish, he claimed, preferred to follow the life of their forefathers in forests and open pastures. In other words, they dissented against rule. Wildness was a refusal to submit to authority.

In *Reveries of the Solitary Walker*, Jean-Jacques Rousseau saw nature as the sanctuary from the traumas of warring and corrupt civilized societies. He resisted the idea of using nature for anything beyond pleasure. The pursuit of knowledge or the technological dominance of nature vulgarized this higher commune with the natural world. He took botanizing for medical purposes as his example. 'Everything that concerns my needs saddens and spoils my thoughts,' he wrote, 'and never have I found true charm in the pleasures of the mind save by completely losing sight of interest in my body.' His book of walks was much loved among sympathetic poets and writers who rose to fame in Europe and North America in the decades after its publication. Those who took his ideas to heart, such as Emerson, Coleridge, Shelley and Wordsworth, made fashionable the perception of wild nature as pure and preferable to the foul excesses of human nature.

But another reason why human wildness was perceived as negative lay in the prevailing fears about which behaviours could be traced back to some innate, animal nature, and which derived from the social environment. From the late nineteenth century onwards, scientists have gradually enhanced understanding of the essential elements that govern the development and function of all living organisms. The experiments of Rosalind Franklin, Raymond Gosling, James Watson and Francis Crick in the 1950s identified DNA, molecules designed for the long-term storage of genetic information, which are the fundamental units of natural selection. Work on the human genome, published in 2001, developed ideas about genes intrinsic to humans in the wilder past that still affected how people lived and behaved. A range of desires and motivations belonged to a time when as a species humans fought to survive in a natural environment from which most were now severed. The Human Genome Project, which began in the early 1980s, was plagued from its outset by philosophical concerns about what truly motivated human behaviour, what control we had over our actions, and how our inherited natures changed us.

At Signy, another of the British Antarctic Survey bases, huge

populations of fur seals were killing off expanses of vegetation that had existed there for thousands of years. Protected
from hunting, their numbers had boomed. Wild animals and
plants altered the harmonies of space and resources; if we too
were wild, why should our wildness be perceived differently?
If we behaved as we were innately patterned to do, destroying
landscapes and their wildlife, why should we curb our behaviour? Would we clatter onwards, plunging headfirst towards
extinction, dragging with us a host of species unfortunate
enough to share their time with us?

Ancient thinkers argued for our superiority by noting that
brute animals had none of the knowledge of nature that afforded them choice of action. Even Rousseau was still be-
guiled by the idea in the eighteenth century. He saw all
animals as ingenious machines that operated to help themselves survive and propagate, and he saw people in the same
way, but with one essential difference. For the brute, he stated,
nature was the sole agent. Other animals could behave only
by instinct. People, on the other hand, could determine their
actions through their reason. But genetic science suggested
that there was less will involved in people's actions, indeed
that our minds made choices before they reached conscious
thought.

In 1962, Vero Wynne-Edwards, a British biologist, published *Animal Dispersion in Relation to Social Behaviour*, a
highly controversial book on the group dynamics of animals,
particularly of long-lived seabirds like albatrosses. Wynne-
Edwards believed that some seabirds adapted to control their

population densities, limiting them to serve the entire colony, the optimum level to give 'the best living to the largest number'. This self-regulation safeguarded food resources from extinction, reducing over-fishing among foraging seabirds that would compete with one another at sea. He pondered whether the flocking of birds was a kind of intrinsic census, whereby a sense of their numbers swept over the creatures as they hung together in the air. If they perceived that the throng overshadowed their resources, the population began to regulate itself through the emergence of non-breeding birds. According to Wynne-Edwards, where individual motives or group motives clashed with one another, 'as they do when the short-term advantage of the individual undermines the future safety of the race', then the needs of the group ought to prevail. The antisocial selfishness of individual birds was supplanted by the need to prevent the decline and extinction of the whole society.

Earlier, another biologist, David Lack, had published *Natural Regulation of Animal Numbers*, which argued that the number of chicks per bird each year was only as large as could be sustained by the natural environment. In the years after Wynne-Edwards' theory was published, Lack became the major opponent of the notion of self-regulation among seabirds, reiterating that the availability of food supplies for breeding birds more closely determined the population outcomes than any other influence, and that animals did not have the capacity to constrain themselves; nature did it for them. The evidence to support Wynne-Edwards' theories remained elusive and his hypothesis was dismissed. Scientists concluded that self-regulation was an unnatural idea, an attractive notion but one not yet found in nature. One could not blame the birds or the seals if on occasion their numbers soared beyond what the environment could sustain, for they had no defence against their own nature.

Perhaps the intrinsic character of our species is less self-willing than earlier generations of thinkers hoped; none the less we have a propensity for self-reflection not possessed by birds. It might be possible for self-regulation to modify the behaviour of a species for the first time, an enlightened course for a sentient animal capable of manufacturing its own sources of nutrition and energy.

The ship left Bird Island during the night and we were now heading for Antarctica. In the early hours of one morning, I went up on to the bridge of the ship. I knocked on the watchtower door and found the room steeped in darkness and quiet. Dougie was the officer on watch, silhouetted against luminous radars, staring out to sea. I crept out of the side door and sat on the bench in the glowering early light, watching the searchlights surprise the companionless icebergs. There was no land in sight any more, but fur seals hooped alongside the ship, hundreds of miles from their shores. It seemed to me, from the colour and gentle nervousness of the ocean, as if we were sailing across the blue body of a whale.

After a while, I returned inside to warm myself up and picked the *History of Antarctic Names* from the shelves. Dougie came over and quietly pointed out his family's names. He was from the Shetland Islands, and several of his relatives had gone to Antarctica before him as sealers and whalers. He spoke of how things had changed in the Shetlands since the exploitation of oil in the 1980s. Now people were richer, he explained, and many newcomers and mainlanders had come to live and raise their families on the islands. But he also talked of how the old culture had rapidly disappeared over the past decade. 'The dialect is going,' he said. 'When you lose your tongue, you forget who you are. You lose everything.'

The sun broke through the clouds. The ocean looked vibrant with possibility, spray and spume gathering momentum like a mind bent on a race. The evening before, we watched humpback whales on the port-side of the ship. Their undersides made the water a ghostly green, flagging their presence before they broke the surface of the sea, up and out of the water in beautiful, rapid arcs, sometimes in union, sometimes in swaying, gambolling play. Their blows were loud, slow and almost worded, as if there was a voice on the wind – at times a puff as good as indignation, at other times a gentle, musical query. They put me into a strange, enervated trance. Now it was the motion of the sea that stunned me, the large, laboured flaunts of the swell. I stood to go outside again, and the pluck of the waves threw me off balance. Outside was the hissing spray, the whistling streams of air. Salt clung to my face and fingers as I held on to the barricades and stared in wonder. The sun shone across the front of the ship like a girdle of rainbow. Occasionally, the spindrift took up these colours,

strewing them across the waves in reds, oranges, greens. Big triumphs of wave shivered to nothing across the bow, and I clasped the storm barriers ever more tightly. Perhaps I should have been frightened before these powerful elements, but I felt only vitality.

The English poet Percy Bysshe Shelley drowned in a wild storm while sailing from Livorno in Italy in 1822. A line from Shakespeare's *The Tempest* decorates his grave in Rome: 'Nothing of him that doth fade but doth suffer a sea change into something rich and strange.' In a letter written that year to her friend Maria Gisborne, Shelley's wife, Mary, described a night when she heard her husband screaming and rushed to his room. Shelley claimed that he wasn't sleeping but was overcome by 'a vision that he saw that had frightened him'.

His vision was of two of his friends by his bedside, half-starved, disfigured and blood-stained, who said to him, 'Get up, Shelley, the sea is flooding the house.' Shelley rose and went to the window, and experienced a terrifying delusion of the sea rushing in.

Ancient civilizations dreaded death by drowning as the snuffing out of the soul into an utter blackness beyond hope, an absolute and endmost extinction. To drown was to fail to transform through death, to fall like slow rain to the sea's floor, with no chance of immortality. The great waters of the world have always threatened to consume civilization. In the mid fourth century, a giant wave destroyed much of Alexandria and its coastline. According to Ammianus Marcellinus, the sea first receded so far that ships were stranded on the seabed. People scurried down to the shore to sift through the beached sea creatures and the grounded ships. 'Thus in the raging conflict of the elements, the face of the earth was changed to reveal wondrous sights.' It was an experience that the Alexandrians ritually remembered for over five hundred years, a reminder that some elemental forces were wholly beyond people's governance. Even in Cornwall, storms were occasionally so violent that the tides broke loose of their informal edges, smashing tethered boats, washing whole houses away. In 1771, a storm carried off a stone pig's trough, returning it after a gap of forty-seven years as a playful condolence during another gale.

In Francis Bacon's *New Atlantis*, his meditation on the ideal state published in 1624, the flood of the fabled land of

Atlantis was a punishment for the overreaching civilization of mankind. Bacon inherited his conceit from the Bible, although the deluge that swallows both men and lands appears in a number of the world's old creation myths. The image of the flood gave shape to the undercurrent of anxiety about the chastising energies of the natural world and the fear that sinfulness on Earth could precipitate cataclysms from which new beginnings would proceed. At the start of the eighteenth century, a gigantic storm washed away a hundred lives on mainland Britain, while thousands perished at sea. People at the time interpreted the event as a punishment from God for the wrongs of people immersed in their material life.

In 1885, the British naturalist Richard Jefferies bound the idea of the flood to humanity's disturbance of the natural environment in his novel *After London*. It opens with 'The Relapse into Barbarism', a vision of society consigned to its watery grave: 'The sites of many villages and towns that anciently existed . . . were concealed by the water and the mud it brought with it. The sedges and reeds that arose completed the work and left nothing visible, so that the mighty buildings of olden days were by these means utterly buried.' Decadent and intemperate, civilization was overcome by the forces of nature, sweeping the clock back to the days when forests ruled the country. 'With wild times,' Jefferies warned, 'the wild habits have returned.' Brambles enclosed fields, the slim lances of grasses and saplings pierced roads – mindless nature was having the last laugh.

Bacon and Jefferies both used the flood as an emblem of

retribution; the scientists undertaking research in Antarctica faced the task of predicting whether future floods would be the fault of humanity. They sought to understand the natural transfigurations of the Earth, to distinguish between natural and man-made phenomena. 'Knowing this natural variability', stated the Scientific Committee on Antarctic Research in 2009, would enable us to judge 'when present day changes exceed the natural state'. Findings confirmed that the amounts of greenhouse gases in the atmosphere were above the natural limit for this time in the Earth's history, largely as a result of burning coal, oil and gas since industrialization. The difficulty for those trying to predict the effects of these changes on the Earth's atmosphere was to deduce which storms and transformations were caused by human influence and which by the Earth's own activities.

In the late nineteenth century, the Scottish scientist James Croll proposed that the irregularities of the Earth's elliptical orbit caused ice sheets to swell or shrink. In the 1970s, a group of scientists began examining the mud and rocks of the deep ocean, searching for signs of oxygen in the shells of minuscule sea creatures, traces that reflected the rhythms of the Earth's uneven orbit. But scientists didn't fully appreciate the intricate relationship of the Earth's orbit and the ice until they began drilling cores from the Arctic and Antarctic ice caps. The air trapped in the ancient ice contained quantities of carbon dioxide and methane whose peaks matched the wobbles of the Earth.

Traces of carbon dioxide in the shells of ancient marine

creatures revealed that there were times when rising levels of carbon dioxide precipitated an ice-free planet and large rises in sea level. Increases in greenhouse gases were natural events but insight into previous warm phases in the Earth's history indicated lower levels of greenhouse gases than those in the atmosphere since industrialization. Our penetration into the planet's frozen and ice-free past confirmed the lethal potential of human activity to match the Earth's powers of transformation. Scientists were forced to conclude that, to fuel industrial growth, humans had extracted and vented huge natural stores of carbon from the Earth, laid down over hundreds of millions of years of the burial of former life. Scientists predicted that if societies continue seeking, extracting and releasing carbon stores into the atmosphere until they are all exhausted, the levels will rapidly rise to those experienced by the Earth tens of millions of years ago, when the world was utterly different.

In imagining this transformed world, the image of the flood returned. In November 2007, while I was on the Southern Ocean, the British economist Nicholas Stern published his warning about the likely effects of carbon dioxide rises caused by our societies. 'Throughout the world there would be serious damage from floods,' he said, 'droughts, storms and sea level rise. At temperature increases of this magnitude, much of the area around the equator would be uninhabitable and there would be a massive movement of population with the ensuing conflict that would result.'

★

As we sailed across the Southern Ocean, I marvelled at the natural splendour, the endless mastery of form. Sea, sky and ice intensified the unplanned harmony of neighbouring shapes. Vast bands of grey lingered across the mountain peaks, creating a twin continent above the snow-lined earth of the Gerlache Strait. Elements mirrored and mocked one another, the icebergs obstinate masses against a landscape of eternally changing light. Hermann Melville described an iceberg as 'a lumbering lubbard loitering slow'. I found them exquisite and it was impossible to tire of them. According to *Polar Oceanography*, a book that I picked from the shelves of the ship's library, 90 per cent of the world's ice clings to Antarctica's frozen continent. Over a thousand miles of iceberg clatters free of the ice shelf each year, endlessly altering for several years until the tangled ocean takes them.

As we began to approach Antarctica proper, the days extended and the light became incandescent. We passed two
cruise ships, the *Fram* and the *Polar Star*, but apart from these
encounters with tourism, we seemed to cross over into another world. I didn't want to sleep. I wanted to concentrate
on the hushed tones of the ice-scape in a devoted reverie. I
stayed on the bridge for hours, gazing across the vast horizon of the ocean, with the icebergs, the incidental islands,
Hoseason, King George, Two Hummock. Minke and humpback whales bounced in and out of the waters, dwarfed by
the sheer depth and expanse of the Southern Ocean. The air
changed; its increasing chill seemed almost to grind itself to
the dust of precious metals. The weather, too, was altering,
heralding the approaching ice edge.

People had become friends during the voyage: Ruth, the
ship's doctor; David, a theatre director on the same invitation
as me; Phil, a tall Yorkshireman, one of the field assistants at
Signy; Paul, a scientist with unbounded enthusiasm for the
soil samples he was studying; and Mandy, whose jester-like
woolly hat frequently disrupted my view of the Antarctic
landscape. I was also friends with members of the crew on
board the *James Clark Ross*: Jerry and Jim, two of the engineers; and Rob, one of the sea cadets whom everyone called
'Gadget'. Together, we shifted between light-heartedness,
playing games on the sticky mats that kept our drinks in
place despite the swell, and moods of amazement and ex-
hilaration at our surroundings. We were always ready to alert
each other to a new discovery on the horizon – a whale, an

albatross, a snowstorm. On the day that we neared the sea ice, I was playing cards in the officers' bar, from time to time glancing up at the view of the ocean ahead. In the distance, I could see a lemon-gold light, a faint signal of the ice world approaching. And, suddenly, we were upon it. All of us dashed off to the foredeck to hear the first, astounding crunch.

On deck, we often talked about the environmental threats that our societies and other species might face in the future. It was easy to associate these waters with impending disaster. Oceans are the ultimate symbol of the Earth's innate destructiveness. But they are formidable forces for survival as well. The deep, dense waters of the Southern Ocean seep up to the surface south of the Polar Front through a wind-driven process called upwelling, releasing the carbon dioxide dissolved in its element. In splendid correspondence, the mid waters that sink on the other side of the front swallow up carbon dioxide. Meanwhile, the Antarctic current, the world's largest, circles the entire globe like a wedding band. It marries the three great ocean basins, carrying salt and heat from one to the other. In the past, over 15 per cent of the annual absorption of carbon dioxide by the Earth's waters has been taken up by the Southern Ocean. But scientists fear that increased upwelling of deep waters infused with carbon along with the parallel soaking up of greater amounts of carbon dioxide into the mid layers of the sea will hinder the ocean's propensity to absorb carbon dioxide in the future, affecting the rate at which the Earth will warm.

Research over the last fifty years suggested that changes to the climate had affected the entire community of Antarctic marine life. Higher levels of carbon dioxide in the waters were gradually making them more acidic, preventing some tiny marine organisms from growing their calcium shells. Diminishing sea ice, especially in the western Antarctic, seemed to reduce the populations of krill, the tiny shrimp-like creatures that feed species of whale, seal, squid and seabirds. In the latter half of the twentieth century, after whalers had slaughtered nearly all of the blue whales in the Southern Ocean, along with humpbacks, fin and southern right whales, scientists expected krill to prosper. However, studies suggested that they had not thrived in the absence of their predators. One possible cause was a decline in populations of phytoplankton as a result of the reduction in the lambent mantle of sea ice that released nutrients for their survival. If the sea ice disappeared entirely, the Southern Ocean would be bereft of a key species, affecting not only whales but penguins and seals. It would become a sea full of vanishings.

Over the next day, we passed into and through the sea ice that fringes Antarctica, with its jagged border and then the beautiful hems of torn and shattered ice stretching like threadbare bridal veils across the freer waters. With their hexagonal patterns, the veils trailed outwards, the colour and density of milk, and glacial ice bobbed on the ripples like dark polished glass. As the ship pushed in further, we passed bright blue pieces of ice, as if lit up by neon, and peculiar pools of sulphuric green and turquoise. We saw an iceberg some fifty kilometres in length, a monster with great grey swathes of cloud and blue crucibles of sky. The black-and-white cape petrels that had followed the ship had disappeared. In their absence, I saw my first snow petrel, white as a new idea. My sighting of the Antarctic continent itself was illusory. I had spent two or three nearly sleepless days on the bridge, while the officers switched from watch to watch to watch and I stared through the midnight sun, waiting for my first glimpse. Finally, I saw a faint silhouette like clouds on the horizon. I said to Dougie, 'That isn't Antarctica, is it?'

'It *is*,' he replied, in his Shetland accent.

I looked through my binoculars and there it was: the ice

shelf ecstatically perfect and the huge, majestic white mountains. From that first sighting, it just kept opening, opening, opening, and with it my understanding that I was approaching something gigantic, a giant's heart regulating the Earth's existence.

On our journey to the British research station, Rothera, we stopped at Vernadsky, a Ukrainian base on the Argentine Islands, off the coast of Graham Land. The oldest operating base in Antarctica, it was sold in the 1990s by the British for £1. The first to visit the area were the members of the 1897–9 Belgian Antarctic Expedition on board the *Belgica*. Under the command of Adrien de Gerlache de Gomery, the ship was originally a Norwegian whaler, its hull made of Norwegian pine, strengthened in greenheart. The first mate of the expedition was the then unknown Roald Amundsen. The tools listed in the ship's inventory betrayed a shift from the vested concerns of the sealers and whalers to those of science. The vessel housed a magnetic theodolite and Von Steerneck pendulum, a toluene thermometer for measuring low temperatures, a cloud atlas, and Le Blanc and Belloc sounding machines for oceanography. For zoology and botany, the ship carried a range of nets, harpoons and hooks, along with hunting rifles, dissecting instruments, and a botanical press for plants. Among the specimens retrieved by the expedition were five new species of mite, twenty-seven mosses, snow petrels, giant petrels, and Weddell and crabeater seals.

In March 1898, ice began to close around the ship, filling

the interstices like swiftly healing wounds. Unable to progress, the men prepared to winter on the ice. The scientists set up astronomical observatories, communicating with the ship by telegraph wires. On 17 May, the sun set and didn't rise again until late July. 'Polar anaemia' inflicted the crew, whose pulses sometimes rose to 150 beats a minute, at other times dropping to under fifty. They began to kill and eat seals, and showed signs of scurvy. One of the sailors died of a heart attack. One experienced fits of hysteria. 'Another, witnessing the pressure of the ice, was smitten with terror and went mad at the spectacle.'

The base at Vernadsky was established in 1947 as the British attempted to secure their claim in the region. Only in the 1980s were modern facilities constructed. The buildings now housed laboratories and offices, a surgery and bathrooms, a balloon-launching building, a skidoo garage, a clothing store, a boiler room, a carpenter's workshop, general stores, kitchens, and reputedly the finest bar in Antarctica. Operated by the Upper Atmosphere and Ice and Climate Divisions of the Ukrainian Antarctic Survey, the base housed wintering scientists who studied ultraviolet radiation, storm effects and plasma irregularities, and long-term changes in the upper atmosphere.

As we approached the base, a huge sign painted with a two-fingered victory V greeted us. We trooped into the building, past a small gym decorated with posters of scantily clad women, and into the bar. There the scientists handed around delicious, homemade vodka and we toasted the polar summer together. On arrival the few women on the expedition were informed of an old tradition that new females on the base should make an offering of their bras. Reluctantly, the other girls conceded, adding their bras to the strings of underwear adorning the room. I was exasperated, but I shook my head in an appearance of good humour and said, 'How do you know I'm wearing one?' Sufficient titillation to keep my underwear in place.

We were travelling to Antarctica during International Polar Year, a global, concerted effort to gather as much information as possible from the continent and its environs. The first of these international scientific efforts took place in 1882–3, followed by another in the 1930s. But the complex of permanent scientific stations was not constructed on the continent until the International Geophysical Year of 1957–8. The two major objectives of the International Geophysical Year were explorations of space and of Antarctica. In 1947, the Russian military fired the first intercontinental rocket, technology which led to the more significant launch of the first satellite. Shot into the atmosphere in 1957 at the peak of the Cold War, Sputnik 1 spooked the American government sufficiently for them to fund a continual vigil by military planes loaded with hydrogen bombs, weapons with a considerably greater destructive capacity than the atomic bombs dropped on Hiroshima and Nagasaki during the Second World War. Sputnik 1 was the first man-made sound from space and anyone with a shortwave radio could eavesdrop on it. There is some evidence that President Eisenhower delayed the launch of an American satellite to allow for relations between the

Soviet Union and the United States to settle down. Whatever the truth, Antarctica staged an important ideological accord when the United States invited the Soviet Union and the other countries participating in the International Geophysical Year to sign the Antarctic Treaty in 1959.

The treaty was a singular document, a diplomatic assertion of unity between nations that had only recently shaken free of war. The treaty prohibited military activity on the continent in perpetuity, including weapons testing, and instituted a series of agreements to safeguard the flora and fauna and the mineral resources of the region. The treaty was a charter of shared values and intentions from the century's terrible wars. In Antarctica, rockets and satellites, and other inventions originating in the violent posturing of those years, were put to more humane use. Satellites enabled the growth of modern oceanography, phenomenally increasing our understanding of the Earth's climate and the expansion and retreat of the sea ice and the ice caps. In the 1960s, the Space Race and the lunar landing swivelled mankind's sights back to the Earth, inspiring some of the environmental movements of the coming decades.

A few weeks before I had left for Antarctica, I was given a tour of the British Antarctic Survey headquarters in Cambridge. After a cup of tea in the 'Icebreaker' cafeteria, I was taken to a chilly laboratory which stored cores of Antarctic ice. Inside these were bubbles of prehistoric air and traces of dust, captured in snowflakes that had fallen over thousands and millions of years. Research into ice cores began in the late 1940s with the work of Danish scientist Willi Dansgaard. In 1954, he proposed that through the frozen annals of ice scientists could establish climatic changes in the past. Engineers adapted technologies used in mining and oil exploration to the conditions of ice and snow, and one of the greatest means of understanding the ancient past became possible. Along with evaluating the composition of the bygone atmosphere, and estimating the global mean temperature from the oxygen isotope ratios, scientists could measure the presence of windborne dust in each layer of ice to distinguish phases of climatic cooling.

I found it a provocative idea that the pristine polar ice cosseted past realities in this way. In 1998, scientists detected pollutants blasted into the world in 1945 in ice-cores drilled from the Agassiz ice cap in the Nunavut territory of the Canadian Arctic. Towards the end of the war, one of the leading scientists working on the atomic bomb communicated to President Truman that they had succeeded in manufacturing 'a new explosive of almost unbelievable destructive power'. As a teenager, this single most horrifying innovation of the war years fascinated me. When I began researching extinction, I decided to go and listen to some testimonies of those who survived the atom bombs of 1945. One late summer day in 2007, not long before I left for Antarctica, I visited the archives at the British Library in London. Entering one of the small listening booths, I spent the day eavesdropping on the wretchedness of the past.

'The raindrops were big and black,' one woman said in

a small, gritty voice. 'What I felt at this moment was that Hiroshima was entirely made up of just three colours – red, orange and brown. The fingertips of those dead bodies caught fire – and the fire gradually burned down. I was so shocked to realize that fingers and bodies could burn like that.'

American forces dropped the first atomic bomb on 6 August and the second on the city of Nagasaki a few days later. Many people died instantly but others inherited the damage from their exposure, from the poisoned atmosphere, through their mother's breast-milk or their father's genetic material; 100,000–150,000 people died by the end of the year. The sound that I heard on the archive recordings was like elastic sheeting being jerked in and out, the noise expanding

and contracting in queer, rhythmic intervals. The invention that destroyed Hiroshima and Nagasaki could obliterate life swiftly and even alter a person's cellular structure, the invisible qualities that made them human. On the tapes, one of the witnesses, Taeko Terumae, born on 19 July 1930, spoke about the days after she began to recover. 'I found a piece of mirror,' she said, 'I looked into it. I had scars just like a mountain range on a map, and my eye like a pomegranate. I almost wished I had died like my sisters. I was so surprised to look like a monster.'

The experiences of those who survived the first nuclear attack were of such incomprehensible proportions that they found it difficult to articulate what they witnessed. One survivor, a photographer, said that the world became bright white as if he'd gone blind. He had his camera with him and, at first, he took a few pictures. But as he neared the centre of the city and saw the mounds of people dead and burned in the middle of their customary activities, his urge to document the reality was completely overwhelmed by his mind's refusal to accept or comprehend the horrors. 'I walked for two to three hours but I couldn't take a single photograph of the central area. Nobody took photos.' Few survivors effectively communicated their experiences of an event of such abnormal suddenness, but in ten thousand years, the polar ice will still speak of these mutant blasts, if melting hasn't hastened the world's amnesia.

The *James Clark Ross* had anchored at Adelaide Island, Antarctica, and I was now on solid ground again, the tipple of the waves of the Southern Ocean only a tremor in my gait. As I walked across the snow, entirely alone, I saw the white heart of Antarctica as a gigantic symbol of memory and forgetfulness, its twin processes of melting and freezing suggestive of what societies preserve and what they consign to oblivion. A few miles from the research station at Rothera, I reached an old caboose hut. It was painted bright red, decorated with black flags that the field assistants used to mark out the hidden crevasses, death traps of contorted crystal, fathomless arteries of swallowed light. My face was white as a geisha's, smeared with sunblock to protect me from the radiation passing through the ozone hole. I stared up at the sky, as snowflakes discoed down.

The planet has mustered enough oxygen to sustain complex life for less than half its history. The evolution of photosynthesis a few billion years ago gradually boosted the oxygen levels in the Earth's atmosphere and gave rise to the ozone layer, which shielded the planet and its inhabitants against the ultraviolet radiation harmful to life. In 1985, scientists from the British Antarctic Survey, Joseph Farman, Brian Gardiner and Johnathan Shanklin, announced that they had discovered the depletion of ozone over Antarctica caused by man-made chlorofluorocarbons (CFCs). Scientists had tested these extensively before use and they were considered both stable and unreactive. But nobody had considered testing what their potential effects might be under the conditions found in the high spring atmosphere over the poles. Used in military planes during the Second World War, and later in refrigeration, air conditioning, aerosols and solvents, CFCs were released into the atmosphere, steadily and invisibly degrading one of the Earth's essential organs for the survival of life. In September 1987, a meeting in Montreal led to an agreement to phase out the manufacture and use of CFCs. Twenty years later, although the ozone hole was stable, it hadn't yet reduced.

I stopped and lay on my back, looking up at the sky. It was so peaceful on the soft snow that I fell asleep for a few minutes, my mind as glazed and unbounded as the great tract of continent before me. When I awoke, the winds had picked up and so I walked back to the caboose hut and sheltered inside. The small cut-out window captured the peaks of distant mountains and the broad, flickering field of ice. Suddenly, it was a purdah of squalling snow. I could see nothing but a dull, agile whiteness. I made a cup of tea and dunked a few chocolate biscuits into the sweet, milky drink, listening to the murmuring winds. I knew nothing of the habits of the weather in Antarctica and wondered how long I would have to stay in the hut. But the storm lasted no more than twenty minutes and then, amazingly, sunlight brought exquisite clarity to that which had been mystery and confusion only moments before. I seized my opportunity and headed back to the station. The wind had etched the crisp veneer of the snow into meaningless calligraphy.

It had come and gone so swiftly, this squall, like a child's tantrum. But I knew that it wasn't benign in the way of the sudden blanketing grey of a rainstorm in London or New York. While I hadn't felt in the least unnerved at the time,

I knew that if I had come up against these unbridled snows while in the landscape, I would have had little to help me survive. Stone blind and chilled to the bone, I would have encountered the natural confines of my animal strength. My body was armoured with skilfully designed and fabricated clothing; the shelter contained a radio which I could use to contact those at the station and ask for assistance and rescue; there were shelves of preserved food to sustain me through the storm; and, once the weather settled, I had a skidoo that could transport me swiftly back to safety. Such equipment and support were improving all the time, as each year more people travelled to remote or dangerous regions of the Earth.

Our actions caused the ozone hole, but the same techno-logical gifts also endowed us with the capacity to foresee the coming catastrophe in time to avert it. They both stemmed from the limitless nature of human thought, people's exhaust-less imagination, languages that indefinitely extended the possible combinations of words and ideas. Was it conceivable that our natural ingenuity and the associated increase in our comprehension of the physical world might balance out the destructive impact of our behaviour? History certainly made this a reassuring possibility. A few years after the end of the Second World War, the American ornithologist William Vogt published *Road to Survival*, a study of the future of human population growth and the world's diminishing natural resources. In the fourth chapter, 'Industrial Man – the Great Illusion',Vogt expressed his fear that societies faced imminent collapse. His studies suggested that, if populations were not curbed at this point in history – over sixty years ago – the human species would soon 'rush down a war-torn slope to a barbarian existence in the blackened rubble'. But Vogt's fears did not materialize, in no small part due to the inventive resilience of our species.

Writing his *View of the United States of America* in the late eighteenth century, the American industrialist Tench Coxe argued that the espousal of technology was natural, because nature itself was but a system operated by the engine of the elements. Coxe and his contemporaries believed new technologies and machinery would eternally 'improve our agriculture, and teach us to explore the fossil and vegetable kingdoms . . . and bring into action the dormant powers of nature and the elements'. There is a good chance that greater technological capabilities will stall the degradation of the natural world and improve the lot of humanity for the foreseeable future. But as environmental constraints on human societies increase, so does the compulsion to separate the natural world from human lives. This severance acts as a shield against the messages of restriction coming to us from nature. Sheltered from the lethal capacity of the storm, my face protected from the damaging radiation from the ozone hole, I could recognize the damage to the environment or I could ignore this and happily while away a few thoughtless hours.

In the weeks of my stay at Rothera, I established my own routine. After a morning walk around the point where the base was built, pausing to study the light as it duplicated the contours of the bergs or to smile at the Adelie penguins as they frisked in the waters, I worked in one of the end rooms at the station. The window overlooked the open sea. As the sun never dropped below the horizon, the light varied in colour between the pale pink of a queen conch's lip and a deep yellow.

In the afternoons, I would go up to the radio room with Tristan, a communications operator from the Orkney Islands, and speak to the various men and women who were out on the ice somewhere, their voices sputtering through blizzards and distance. Tristan and I formed a double act, making jokes and trying to boost the morale of those far from home on the ice. The field scientists spoke of the beauty and ferocity of their surroundings, of their passion for the research they undertook at these remote locations, and of the tedium of days spent cramped inside their tents waiting for a lull in the tyrannous wind. More than anything, they spoke of their families. They wanted company, and while they appreciated the time we spent on these brief daily chances of contact, it was sons and daughters whose birthdays came and went, or their closest confidants, their husbands, wives, lovers, that they mentioned most wistfully.

I, too, missed those closest to me. But I longed for something less tangible as well. The growing sense of my inability to survive in this environment without exceptional facilities and gadgets made me yearn for a greater accord with the landscape. I longed for fruiting trees, for rivers and shorelines, and for a place where I could sustain myself without such sophisticated technology. Since the latter half of the twentieth century, better equipment and support have markedly lessened the risk of death in Antarctica. Even the deaths of Tom Allan and John Noel of the British Antarctic Survey in 1966 were highly unlikely to happen in the twenty-first century. The pair froze to death in the stern winds that rushed down the plateau to where the men huddled by their snow hole. Fresh snow also killed and buried the dog teams and sleds with which they had journeyed. I wondered what they thought about as the blood in their veins slowed, stiffened and crystallized, as if Antarctica was remaking them in its own image – whether they, too, thought solely of home.

I began to feel anxious that I had no landscape to which I was truly native, gifted with an intrinsic knowledge of its subtleties and tempers. It was a bewildering homesickness for something I had not yet possessed, a homesickness that arose not from memory but from the want of it.

During my last days in Antarctica I thought more and more about the resolve to avoid fatal and restricting attachments to specific localities with finite opportunities. What would it matter to people if landscapes lost their distinctness and became merely reflections of other places, if this variety didn't determine our survival? Most other animals were still coupled to their specific habitat or ways of propagating, and would not survive the many revolutions to come. An amazing creature evolved to live in the chilling waters around Antarctica, the Antarctic toothfish, *Dissostichus mawsoni*, one of the so-called ice fish. In the depths of the ocean, without certain minerals, the bones of the species altered its body density for different parts of the Southern Ocean. Antifreeze slugged in its veins and its spleen grew to absorb the ice crystals from its blood. Such staggering resilience progressed with the infinitesimally slow cooling of the Antarctic region. If Antarctica swiftly warmed, these elegant beasts wouldn't escape the sea change.

6

Savages

I left Adelaide Island a few days before Christmas, flying to the Falkland Islands from the runway at Rothera. As we lifted into the blue sky drenched in gold by the constant light, I watched one of the *James Clark Ross*'s lifeboats drilling on the still water near the base. The boat left small, unremarkable triangles in its wake like a water boatman on a pond. Rising far above Rothera, the human presence was negligible against the sightless white continent. It was difficult to comprehend that our species could threaten this monumental volume of ice. Everything began to shrink as in a fairytale, toolboxes scattered across the gritted concourse, construction vehicles with their scoops, cranes and claws. Thick black power cables strewn everywhere became fallen hairs, the yellow control tower the stump of a reed. One of the orange twin-otter planes took off as if a child had released the string on a balloon. A hundred or so oil drums became a clutch of smooth, round pebbles deposited by chance.

In a month or so, I would begin my journey back to England. I was by now convinced that people would dream up discoveries in an effort to circumvent the checks of nature for as long as possible, but it appalled me to think that we might have knowingly constructed new civilizations in ways that involved the death of thousands of other species and their landscapes. It troubled me further to recognize that indigenous peoples and their cultures have also become extinct in the course of advancement. Before I left, an unplanned excursion to Keppel Island provided me with a vivid example of this. In the Falkland Islands, I stayed in a small white bed and breakfast run by Celia. Its frontage was a glass conservatory, a pocket paradise of roses and tomato plants in the midst of the peaty moorland, bleak but for bushes of yellow gorse and sprays of diddle-dee berries. Celia and I often sat there together, enjoying the sunshine and one of her cakes, nattering over cups of tea. One day she began telling me of an outlying island, little visited now, owned by an Englishman in his nineties. 'They had the old mission there,' she said. 'You know, for converting the natives from Patagonia.'

As she spoke, she triggered a string of memories from years earlier. I recalled walking up to the village of Paul in West Penwith with my father as a great storm blew in from the sea. Grey fragments of an ancient stone cross cursed the elements. We both scuttled along as lightning revealed a church. We pushed on the oak door and were inside. My father sat on one of the pews, reading a pamphlet, while I stared out through the rain at the distant headstones. Curnow. Hockin. Cattran. Jacka. Carne. Richards. Grenfell. Kelynack. Morrish. Snell. Jilbard. Maddern. Bodinar. Keigwin. Cornish. Pentreath. Rouffignac. Kneebone. Rosewarne. Hundreds of names. The years they were born and the year that they all died, during the cholera epidemic of 1849.

Retracing my steps past the pulpit and the choir stalls, I noticed something we'd missed on our hurried entrance: a memorial on the south-facing wall. Inscribed to 'The Three Johns', it told of a trio of fishermen from the village who perished in Tierra del Fuego while attempting to convert the natives to Christianity. The story instantly captivated me.

Reading dozens of accounts, I went on to discover that at noon on 7 September 1850, a party of seven missionaries under Captain Allen Gardiner, including the Cornish fishermen John Badcock, John Pierce and John Bryant, set sail on board the *Ocean Queen*. The party skirted the Cape Verde Islands and Ascension Island, finally arriving at the storm-cuffed shores of Picton Island in Tierra del Fuego. In an abysmal verse, Captain Gardiner wrote, 'Wild scenes and wilder men are here.' Over the following months, the missionaries endeavoured to establish themselves and pursue their goals, while the physical environment and the natives remained resolutely against them. Those who financed the mission fatally overestimated the ability of the missionaries to survive in such an unfamiliar and unpitying landscape. It never occurred to them that the men might struggle to eat. Captain Gardiner sent frequent assurances that the missionaries on Picton Island could easily hunt for wildfowl and fish.

As time went on with no news, those at home became concerned. By October of the following year, several ships were on the lookout for the missionaries. Captain Smyley, whose schooner, the *John Davison*, was one of the vessels engaged in the search, recorded the following in his diary on 21 October 1851: 'Came to in Banner Cove, Picton Island. Saw painted on the rocks at the entrance of the Cove, "Gone to Spaniard Harbour". Went on shore and found a letter written by Captain Gardiner saying, "The Indians being so hostile here, we have gone to Spaniard Harbour."' Smyley ordered the boat onwards in the belief that they would find the men alive and safe. But the following day, he described what met them on their arrival. It was blowing a severe gale. They struggled to steer the boat ashore but once they finally managed the run into Spaniard Harbour, they caught sight of a wrecked boat on the beach. A dead man lay inside like a question-mark. One of the crew cut off the name sewn into his frock, 'Pierce'. They found a grave marked 'John Badcock', and then a corpse washed to pieces by the rain. 'The sight was awful in the extreme,' wrote Smyley. 'By the journal, I find they were out of provisions on June 22, and almost consumed by scurvy.'

The Cornishmen perished alongside one another, the first of the mission party to succumb. They had all earned their place on the voyage by merit of their prodigious abilities to sail the Atlantic Ocean and feed themselves and their people from the fish of the sea. At Tierra del Fuego, where the Atlantic meets the Pacific Ocean and fierce winds and giant waves had terrorized sailors for centuries, many ships trading in gold and wool had sunk with the bodies of other ambitious men. Here, in unfamiliar territory, the Cornish fishermen forsook their traditional knowledge of the sea and starved to death.

'They might have continued to live in every comfort in their own country,' wrote Robert Young in *From Cape Horn to Panama*, a contemporary account of the mission, but instead they attempted to settle in one of the bleakest regions in the world, sleeping outside in the wind and hail, wet, cold and hungry, trying to survive on wild celery, mussel broth, limpets, and boiled seaweed from the rocks.

I was fascinated by the era that gave rise to these events. The cholera field in Paul churchyard was a stark reminder of the poverty, filth and restrictions of many lives at the time. During these years, the technology of the railways was rapidly spreading around the world, the radical social ideas of Marx, Engels and John Stuart Mill were diffusing across Europe in the wake of the revolutions of 1848, early photographic techniques were being refined, undersea telegraph cables were making it possible to communicate across continents, and in 1876 Alexander Bell would patent the telephone. It was in this era that Charles Darwin's far-reaching explanations for the origins of diverse life on Earth appeared.

I'd considered the extinction of both places and animals, and also the retreat of native knowledge among some generations. Now Celia's talk of the mission island, which revived my childhood interest in the Three Johns and their efforts to subdue the natives of Tierra del Fuego, inspired me to explore extinctions of peoples and their societies. The story pointed towards devouring, dominant cultures, where people

sought uniformity and a deliberate suppression of nature. I
wanted to seek the lineage of their convictions.

On Keppel Island a mission was constructed following the
failure of Captain Gardiner's expedition. Gardiner starved to
death, but according to his diaries, he died a 'happy death'. A
short while before, he wrote a series of recommendations for
subsequent missions: 'During the progress of their education,
and acquiring and teaching language, the natives should be
employed in tilling small plots of garden ground, and also in
tending stock on the Mission grazing farm.' Those that followed him should transfer the station to East Falkland, he
said, with a few natives from Picton Island who might teach
the missionaries their language. The natives should learn
English, so that they might become skilled instructors too.
'They are exceedingly quick in imitating sounds,' Gardiner
added. Only a few years later, the missionary ship named
after him, the *Allen Gardiner*, was ready to set sail for Keppel Island, part of the recently claimed British colony of the
Falklands, an archipelago of untamed nature. Here, the South
American Missionary Society attempted to fulfil Gardiner's
vision, shepherding natives from Tierra del Fuego to the island, instructing them in Bible English, and converting them
to Christianity. The Reverend George Packenham Despard
of Redlands, Bristol, held the office of superintendent of the
mission, along with his wife and their young charge, Thomas
Bridges.

A desire to visit Keppel island consumed me. Celia helped
me contact a representative of the old gentleman in England.

With a bag full of coins, I used a public telephone and received confirmation from a shaky, male voice that I was welcome to explore the island, as long as I didn't stay the night. 'There's nothing and nobody there,' he warned me.

In the sunlight, the waters around Keppel Island were livid with kelp. There was a stone jetty on the north shore, once used by the missionary schooner, the *Allen Gardiner*. It was here that a Falkland Islander called Allen Whyte steered his boat, allowing me to alight. I stood quite still and watched as the boat returned to sea, receding to an idea on the horizon. I was alone. The forecast had warned of gales, but the seas were gentle, the cloudless skies an oath. I turned towards the sun that beat its light stubbornly on the tussock grass and climbed up from the shore, past troops of steamerducks and upland geese. In the distance, I saw the abandoned settlement. As I neared it, I became aware of a strange chattering resisting the silence. I hesitated, trying to find the direction of the sound, and then began to walk towards it. I passed a small enclosure of light-grey graves, gilt with lichen. TO THE MEMORY OF CHARLES HENRY BARTLETT. BORN DEC 30TH 1863. DIED MAY 30TH 1864. OF SUCH IS THE KINGDOM OF HEAVEN. What were these fellow countrymen of mine doing in this place? Did they relinquish their homeland, or did memories of it cloud their vision of their new landscape? Knowing a landscape brought a person back to the sobering limitations and realities of their own nature. Otherwise, loosed from specific attachment to a setting or a way of life, characters and attributes

could become quite altered and strange. The graves pointed towards the boarded-up houses, and, with keener sight, I saw that shadows marked other sunken burials on the hillside. Yet the place wasn't gloomy. Indeed, I felt thrillingly free.

The chattering increased; I followed. In the valley below, I prised apart the undergrowth of wild flowers and shrubs and saw what lay sheltered: the memory of a garden, slit by a stream now dry as dust. It made me think of one of the books about the missionaries that I had read many years ago, which described a plantation where dark hands once tended fruit and potatoes, where southern beech and Monterey cypress colonized foreign soil.

I squinted through the shadows. Yes! There were the cypresses, standing tall amid the endeavour. This was the place – a little plantation grown to provide food and income for the missionaries and their subjects. The chattering was close by now: a single bird sitting in the highest branches of a tree, bibbed in red and black-beaked. I racked my mind for fauna recently learned from the guidebooks to the region. A red-backed hawk? The crested caracara? I didn't know. Beyond the bird, in the space into which its song flowed, there was the stone rubble of two further dwellings. No sign recorded the names of their prior occupants. Perhaps the names themselves belied their origins, baptisms made senseless by death, for only in life can a person's background be turned against them.

Turning upslope, I spotted a handsome, three-gabled house, conspicuous in its salvation, extended into a dairy. In its shadow were the ruins of a cottage. From an old map, I learned that this was once occupied by Ookokko, one of the Fuegian Indians, the building humiliated through the years into nothing but the spectre of a hearth. And beyond this, more recent remains of a red-brick shearing shed, added to cater for the three thousand sheep that the mission kept towards the turn of the century; a 1950s farmhouse with flat tin cladding and the characteristic Falkland Islands sou'wester roof; and a green vehicle steered to deadlock by the gorse growing in its fuselage. The garden, the farm: unruly earth and flora tamed to yield a profit, each act of cultivation intimately

associated with the idea of possession and, by extension, with the augmentation of one's individuality, one's own interests. To possess and tend land became the means of supplanting any original identity.

The mission had a distinguished history. In 1831, while still living in Cambridge, Charles Darwin received a letter from one of his colleagues, John Stevens Henslow, recommending him for a survey of the southern extremity of America. Darwin accepted the challenge and the subsequent voyage of the *Beagle* took him from the Falkland Islands to the Galapagos and beyond, skirting the length of Chile, in and out of the maze of islets fringing Patagonia and Tierra del Fuego. These distinctive landscapes were the backdrop to his theories of extinction and evolution.

On 17 December 1832, the *Beagle* anchored at the Bay of Good Success, where Darwin encountered the natives of Tierra del Fuego for the first time. 'It was without exception the most curious and interesting spectacle I ever beheld,' he wrote. 'I could not have believed how wide was the difference between savage and civilized man: it is greater than between a wild and domesticated animal, inasmuch as in man there is a greater power of improvement.' He considered these people utterly pitiful, noting in his journal that the poor wretches were stunted in their growth and that their skins were filthy, their hair tangled.

Darwin's expectations of the people of Patagonia had been heightened during the months of the voyage by the young Fuegian Orundellico, known as Jemmy Button, who was returning home after a sojourn in England. On an earlier voyage to Patagonia, Vice-Admiral Robert Fitzroy, the captain of the *Beagle*, had persuaded Orundellico to board his ship in exchange for a mother-of-pearl button. In England he became the source of fascination for Queen Victoria and her subjects, and the focus of experiments in civilizing 'primitive' people. Darwin portrayed him as a vain man who would become agitated if anything dirtied his highly polished shoes. Jemmy adopted English colloquialisms like 'Too much skylark', and only occasionally revealed the keener senses that Darwin considered his native attributes, such as his exceptional eyesight, envied by the sailors. After a short navigation around the inlets of Tierra del Fuego, they returned Jemmy to his tribe. To Darwin and the others on the *Beagle*, he appeared to have forgotten much of his native language and to be ashamed of his countrymen. Darwin commented that it was both laughable and pitiable that Jemmy tried to speak to his own people in English. Yet, once in places well known to him, Jemmy reputedly guided the boats through to safe anchor and recognized each islet and point by its native name.

On 5 March 1834, the *Beagle* crew returned to the safe anchorage where they had left Jemmy. Shortly afterwards, a canoe appeared, with a small flag fluttering from its bow. One of the occupants was busy scrubbing the traditional paint from his face. It was Jemmy, thin, bedraggled and naked but for a scrap of cloth about his waist. 'He was ashamed of himself,' said Darwin, 'and he turned his back to the ship.'

In 1858, Jemmy spent six months on Keppel Island. In that

time, he managed to persuade a further three Fuegians and their wives to join the mission. One of them was Ookokko, later baptized as George. On Christmas Eve, he and his wife had a child, whom they named Shukukurtumagoon, meaning 'Son from a House Thatched with Grass'. In the 1860s, missionaries took Wamestriggins, whom they renamed Threeboy (from Jemmy's garbled English for his third son), to England; he did not survive the return voyage. According to witnesses at the time, his last words were, 'I believe in God the Father Almighty,' before dying on the no-man's-land of the ocean. Changing the Fuegians' names, as the missionaries did when converting them, as well as relocating them to another country, helped to sever them from personal circumstance and place.

I peeped through the windows of the 1950s farmhouse, the glass opaque with dust, and saw the poised rose-patterned curtains, the book on the bedside table, angled as if to be taken up again, the tender irony of its title, *Dust of Life*. In the kitchen, there was a capless bottle of Fairy Liquid, an opened packet of Weetabix, and an ancient-looking Rayburn stove over which towels draped, in suspense of heat. I wrenched open the outhouse, where rusting tin buckets still held the pungency of peat, and a small white-and-green box of Lamb Tonic, 'for weakly or convalescing lambs', was reminiscent of a time of fertility. The farmhouse stood on the foundations of an earlier residence, where Thomas Bridges lived, the first European who, in 1871, established his residence on Tierra del Fuego and compiled the Anglo-Yahgan dictionary from the Fuegians' native language. Apparently, Bridges' descendants still farmed the land at Harberton, in Patagonia, that he cultivated in 1886. Keppel Island was only ever an interim measure, a bridgehead to the establishment of a mission in Tierra del Fuego. Ultimately, the aim was to persuade the natives to convert one another in their own homeland. I hiked up past the shearing shed, which bore the outline of a human head embossed with the initials SAMS – South American Missionary Society. I wanted to reach the top of the hill and to sit in tranquillity on the tussock grass, surveying the ruins below me. It felt exhilarating and a little frightening to be alone like this, wholly alone in a landscape that had haunted my imagination ever since I had hurried into Paul church with my father to shelter from that rainstorm and become fascinated by the mystery of the three fishermen from a tiny village in West Penwith.

It seemed to me now that this story resonated for reasons other than the eventual, perhaps inevitable, demise of the Fuegians. Certainly, a slow kind of death seeped into the lives of those the missionaries hoped to enlighten. The extinction of the Fuegians and other indigenous people was brought about by a culture that rationalized ever greater distances between society and nature. Proselytizing was an essential method of imposing social codes on large numbers of people, but it also legitimized the invasion of one culture by another deemed to be more effective. Ancient philosophers, such as the Greek Stoics, sought to divide humanity from the rest of nature by virtue of the inexplicable, rational capabilities of

the human mind. The concept of the *pneuma* was channelled
into the Christian faith as the human soul, and people came
to believe that pursuing the subjugation of the natural world
for the convenience of our species was an act of goodness in
a designed universe. Yet the missionary zeal betrayed a hidden
fear of humans whose lives seemed contiguous with nature.
Swift proselytizing or else suppression was needed to avoid
challenging long-standing and profitable beliefs.

The missionary experiments in Tierra del Fuego were an
unmitigated disaster. The Fuegians remained resistant to sur-
rendering their traditional ways of life and their territories.
Of the prehistoric people of Patagonia, some were seafaring
canoeists in the archipelagos of Tierra del Fuego and others
inhabited the mainland as hunters. Over time, they diversified into the Tehuelche Indians of the pampas, the Onas of
Tierra del Fuego, the Kawesqar and the Yahgan tribes. The
Tehuelche predominantly hunted a kind of llama called the
guanaco. During the seventeenth century, they began using
horses introduced by the Spanish, which enabled them to
travel longer distances to trade guanaco skins with colonists
and mariners. Despite what they gained from the Spaniards,
contact with the foreigners transmitted diseases and alcohol
abuse spread among them, which eventually led to their extinction. In the latter half of the nineteenth century, the Chilean government encouraged the immigration of European
colonists. As the colonizers and Chileans found some areas of
Tierra del Fuego suitable for the grazing of sheep, they set
about clearing the land. A report in the *Daily News* noted

unsentimentally that they undertook this conversion with 'the single inconvenience of the manifest necessity of exterminating the Fuegians'. The colonists systematically and aggressively targeted the Onas, who were either murdered or placed in sanctuaries run by Spanish missionaries. From a population of several thousand in 1870, their numbers had collapsed to less than a hundred by the early twentieth century. The Kawesqar, who fished and hunted sea lions, and smeared their bodies in animal fat and soil to protect themselves from exposure to the harsh environment, also succumbed to alcoholism and disease after association with European sealers and traders from Europe and America. Reputedly, three descendants were living at the close of the twentieth century, the last of their people. The Yahgan, the tribe to which Jemmy Button belonged, were nearly extinct within a century of contact with the missionaries, as a result of syphilis, measles, tuberculosis and fighting. Although several thousand people claimed in the 2001 census to have some Fuegian ancestry, missionaries and colonists did exterminate, whether deliberately or unintentionally, through encroachment on their territories, the peoples of these various tribes.

Towards the end of the nineteenth century, as it became increasingly apparent to the missionaries still working with the natives of Tierra del Fuego that their extinction was imminent and probably unavoidable, outsiders became concerned to preserve and trade what was left of the dying cultures. One of the priests from the Spanish mission, Maggiorino Borgatello, established a regional museum in 1893, gathering ethnographic materials from the tribes for permanent display. Meanwhile, major museums in Europe began to demand artefacts from the country, and another of the priests, Angel Benore, undertook the task of collecting and sending items overseas. It was a nostalgic response of little effect among those convinced of their beliefs.

The sun had dipped towards the sea and it would soon be time to leave Keppel Island. I wandered down the hill again to reach the coastline, watching the harping oystercatchers as they picked over the old fish weir that the receding tide had exposed. I looked back at the settlement, benign in its ruination. Mr Whyte had told me to expect him by four o'clock but I had forgotten my watch. It struck me that it was at least an hour or so beyond our agreed meeting. Had he come while I had been so absorbed in my own reflections on the hill that I had missed him? Or had something happened to him out at sea? Nobody else knew that I was on the island and I had no means of communicating with anyone. I began to feel scared and started to worry about how I might survive if he didn't return. I could find a stream, I told myself. I could break into the farmhouse and shelter inside. I could cut the branches of the trees for firewood. I could collect berries, kill the geese, strike open mussels that I found stranded on the flats of the beach. I would create a huge bonfire on a clear day and hope that the smoke signal would rouse someone with eagle eyes on a nearby island to rescue me. But, of course, within the hour, I heard the faint but unmistakable sound of an engine and I ran round to the point where Mr

Whyte was now idling his boat. As I stepped aboard, he explained that he had come earlier but that the tides had made it impossible for him to approach the shore and he had gone away until more favourable conditions allowed him to enter the bay. As we dashed out across the waves, I felt foolish at having been consumed with thoughts of survival after such a brief time alone on the island, and smiled to myself at the appropriateness of the name of a narrow strait between two islets that Mr Whyte steered us through: Anxious Passage.

Until the theory of evolution, thinkers across the centuries dreamed up contrary ideas to explain the history of humanity, from the fall from an age of innocence or sublimity to a world governed by inevitability and recurring cycles. The image of man's progressive perfectibility retained its magnetic appeal. How such perfection might arise was far more contentious. Success depended on a belief in the power of education and exposure to civilized ways of life to transform human nature.

This picture was consistent with long-standing ideas about the transfiguring potential of cultural refinement. As Saint Augustine imagined in *City of God*, the education of the human race progressed through the ages, eventually rising from earthly to heavenly things. In *An Essay Concerning Human Understanding*, John Locke's image of the mind as white tablet suggested that humans gathered their traits and habits through the ascriptions of time and experience on an otherwise unwritten nature. A century before the publication of Darwin's ideas, the French economist Jacques Turgot presented his *Notes of Universal History*, in which he envisioned culture as one of the chief instruments of progress. 'Through the advance of languages, of physics, of morals, manners, of sciences and arts,' he said, 'people marched towards perfection.'

Societies that were independent of the cultures which gave rise to these ideas were thought of as crude and dissolute. *Sylvestres homines*, the 'wild men' described by Cicero, were neither Greek nor Roman and no better than beasts. They were wretches outside the dominant culture, feckless and unformed. This old prejudice found another influential form in dozens of books that circulated in Europe from the early Middle Ages onwards, which argued that the spread of Christianity divided men into those who had embraced the religion and those whose lack of Christianity left them in a more primitive state, ruled by baser forces.

These ideas mingled with the concept of *Homo ferus*, identified by Linnaeus in his *Systema Naturae* of 1758, as a human suspended between civilized man and the lower orders of animals. This category of humanity was applied to children like the Wild Boy of Aveyron, who was discovered in the late eighteenth century, scarred, speechless and naked, in some woods in France, where purportedly he had lived alone since infancy. The failure of these 'wild' children to acquire language skills touched latent anxieties about how close humans were, especially in youth, to earlier bestial natures. In 1767, Adam Ferguson argued in his *Essay on the History of Civil Society* that the individual, in every age, ran the same race from infancy to manhood. Every infant, or ignorant person, he said, is a model of what man was in his original state. The tacit question in these works was the nature of this original state: was it one of goodness or of depravity?

The work of the mission on Keppel Island was driven by ideologies that sanctioned the suppression of nature and drove a wedge between people's lives and the natural world. In agricultural Europe, notions of self and the ownership of land were inseparable – the name on the deeds of owned and tended land. But the nomadic hunters of Patagonia, despite the skills and adaptations that allowed them to survive for thousands of years, by turns hunting, gathering or scavenging, belonged to their landscape quite differently, living at the mercy of nature rather than trying to subdue it. The missionaries and those who sent them hoped that by bringing agricultural and social skills to the Fuegians, along with the means to begin to dominate the environment, they would also create new civilized identities among the people whom they saw as weaker and more barbarous. They believed that while human nature in its aboriginal state might be ungovernable at first, the strong hands of faith, law and education would overcome these defective origins.

There is a different quality to thinking while at sea for extended periods. The action of the swell unsteadies the body, the eye scans a limitless horizon with no outcrops on which the mind might anchor, and the infinite, directionless movement of the ocean untames both vision and thought. On the *Beagle* voyage, Darwin began to understand that living things were freighted through time by inherited information, behaviours and physical adaptations that reproduction could pass on to the next generation. By candlelight, the storm barriers in place, the materials of the ship pitching against the waves, Darwin chased down his radical ideas. He acknowledged that the greatest means of keeping animal populations in check was nature itself, the raw forces of earth and atmosphere. To escape such restrictions or the likelihood of extinction, a species would have to outwit nature. Thinking about the Fuegians, Darwin believed that natural and hereditary effects had fitted them for both the climate and resources of their country, but that the harsh environment severely curtailed their numbers.

The concept of progressiveness in Darwin's evolutionary theory seemed to reflect earlier presumptions about the stages of refinement that raised mankind above the ranks of the animals to the beau ideal of creation. In the early twentieth century, the evolutionary biologist Julian Huxley linked the emergence of dominant species with the attainment of the greatest complexity. The most salient fact in the evolutionary history of life, he said, was the succession of what palaeontologists called dominant types. These were characterized chiefly by complexity, and their spread caused the partial or even total extinction of those in competition with them. In accordance with the theory of evolution, human beings had escaped extinction to become the dominant type by refining their natural form.

For his part, Darwin recognized that the extinction of groups of humans had taken place in the past, and that injurious or intrusive alterations to the way of life or habitat of any people closely bound to nature put them at risk of reduction and ultimately extinction. In applying his ideas of natural selection to humans as well as to animals, Darwin in fact dethroned humanity from its pre-eminence over nature. Darwin argued that natural processes could explain some of the properties of the human species and that these were no different from those which gave rise to other life forms. Blurring the distinction between humanity and the rest of nature, his theories stripped human societies of their superior rights to the Earth's resources. An emphasis on the powers of civilization allowed industrialized societies to avoid Darwin's conclusion for a time. When missionary zeal declined, they continued to justify their dominance and destruction of the natural world by claiming that too close an association with nature was backward.

In the early twentieth century, the anthropologist Charles Wellington Furlong studied the Fuegian tribes and concluded that their relative weakness had driven some of them to the remotest and wildest stretches of the country and that this predisposed them to their eventual demise. The harsher the environment in which the different tribes lived, the more physically demeaned such people appeared, their stature stunted, their legs bowed. 'The more I have seen of these and other primitive tribes,' he stated, 'the more I am convinced of the direct and potent influences of environment.' In the

1960s, an American anthropologist, Dr Hammel, found that the Fuegians could sleep in temperatures of 0–5°C with little clothing, while their base metabolic rate remained high and heat still reached their extremities. Hammel concluded that this adaptation, together with layers of fat and their 'stout ankles and squat, short-toed feet', allowed the Fuegians to survive and forage in chilling temperatures. The traditional mode of life of all the Fuegian tribes, he summarized, was closely related to their natural resources, and had led to a physiological adaptation as well. Those belonging to industrialized countries could level the perceived barbarism and ultimately the demise of the natives of Tierra del Fuego at the inherent fragility of their bodies and society, which had failed to control and gain independence from their native landscape.

Darwin noted that the Fuegians were excellent mimics, a trait he ascribed to senses sharpened by survival against the elements. In the wilder past, it would have been a safer strategy to mimic others. Gambling with something untried under dangerous conditions could hasten death. But imitation would only remain of benefit in circumstances that stayed more or less the same. If people did nothing but imitate one another, they would eventually lose sight of the changes in their environment and, before long, these changes would render the imitated behaviours redundant. And so the ability to innovate was also indispensable to survival on the mercurial Earth. As such, the European compulsion to innovate and the Fuegian resistance to change were both natural strategies to which either culture might have recourse. One was driven by limited resources and the other by survival in an environment less speedily altered by the smaller population.

Readers of Darwin cherry-picked his theories to argue for the general course of mankind towards improvement, but other ideas introduced some doubts. Darwin's propositions agitated old terrors both about the power of nature and about the possible unruliness of human nature. Was our inherent fallibility the source of the degeneracy of the civilized world? Rousseau had argued that the failure of humans to perceive the goodness of their true nature led to corruption during the later stages of civilization. In his *Essay on the History of Civil Society*, Adam Ferguson included a chapter on the corruption of 'polished nations', claiming that while one might not wish to relinquish favourable progress for some imagined earlier stage of innocence, too much civilization was no more than an extreme expression of our selfish animal natures. 'Whole bodies of men are sometimes infected with an epidemical weakness of the head, or corruption of heart, by which they become unfit for the stations they occupy, and threaten the states they compose, however flourishing, with a prospect of decay, and of ruin.' Ferguson put this down to the uncontrollable selfishness intrinsic to human and animal natures. Turning the concept of human progress on its head, Ferguson believed that man's capacity for advance was simply the manifestation of a trait shared by all animals to utilize any characteristic or faculty that might encourage their own survival and expansion. In some cases the pursuit of perfection would be benign or even beneficial, but in others it could be damaging or pernicious, as the competition among humans would cause some to indulge their own desires and appetites at the expense of others.

From one angle, Ferguson's ideas supported the favoured hypothesis that education and society formed human nature, agreeing with Rousseau that corrupt civilization merely warped the better aspects. But Ferguson's theory hinted at something far more disturbing, the possibility that the destructiveness of mankind was due to an inability, even within civilized society, to escape human nature, which was profligate in its origins and to which people could suddenly and unavoidably revert.

The concept of the struggle for life reassured societies that strength and canniness would allow them to triumph, but Darwin's theories were far more subtle than a simple narrative of progress; they exposed the complicated balance of conditions and processes implicated in the growth and diversity of life forms. One aspect of this complexity that concerned Darwin but was often ignored subsequently was the retention of physical and behavioural traits through a process he called 'reversion'. While many species were highly adapted to their habitats, studies of the skeletons and anatomies of various animals revealed shrunken or vestigial organs, the legacy of earlier habits. Darwin postulated that these 'rudimentary parts' might respond to 'the various laws of growth, to the effects of long-continued disuse, and to the tendency to reversion' – the hidden readiness to return to lost ancestral characteristics, endlessly passed down through the generations.

The curious appearance of vestigial limbs in the skeletons of animals gave the naturalist clues as to the former conditions in which their ancestors lived. Darwin took as his example the whale, whose foetuses seemed to show signs of teeth that were absent in the jaws of adults. As far as Darwin could fathom, the inherited ability to revert to an organ or trait once used by the animal could be invaluable, especially in guarding against the contraction of opportunities and a reduction in adaptability as the creature tended towards greater specialization. While this concept consumed Darwin's thoughts, it was less reassuring than a straightforward vision of linear progress. It was a challenging notion of the effects of the past that endured both inside and outside people's bodies and over which we were powerless.

What might happen if human civilization were destroyed? Would people find themselves in a purgatorial stasis, unable to express their desire for new opportunities and growth, or would such times slowly reduce the human species once more to the grubbing nature of beasts? Early thinkers emphasized the uniqueness of the human ability to walk upright, but Charles Lyell proposed a more disturbing possibility. The staggering and falling of children, he feared, were evidence of a latent susceptibility in humans to return to the quadrupedal state.

Darwin lent weight to this grim possibility of resurgence to an earlier wildness and bestiality. 'Injurious characters,' he wrote, 'tend to reappear through reversion, such as blackness in sheep; and with mankind some of the worst dispositions, which occasionally without any assignable cause make their appearance in families, may perhaps be reversions to a savage state.' As he made his case for the evolution of mankind, Darwin began with the anatomical discovery of a particular muscle in the feet of man, also present in apes. He went on to cite the scattered hairiness, particularly of the male physique, the remnants of the furry hide of beasts. Even more unsettling to the straightforward perception of the perfected body of mankind were the canines, which Darwin argued demonstrated the potential for a return to savageness.

While this was feared, the so-called 'savages' encountered by European explorers – the peoples of non-industrialized countries – heightened the perceived superiority of progressive and industrial societies. Despite Rousseau's vision of the savage as noble, these people were more usually seen as evidence of the brutish nature of humans still forced to battle for survival against nature, confirming the need to diminish any potential reversion to a wilder or former state. Central to this was the ability to obtain mastery over the natural environment. 'Man can resist with impunity the greatest diversities of climate and other changes,' Darwin wrote, 'but this is true only of the civilized races. Man in his wild condition seems to be in this respect almost as susceptible as his nearest allies, the anthropoid apes, which have never yet survived long, when removed from their native country.'

Before arriving in Tierra del Fuego, the *Beagle* anchored

in Buenos Aires. Darwin's observations of Argentina's native inhabitants signalled their endangered condition, forced out from their previous habitations to roam hopelessly, shrinking towards depravity. Not only had whole tribes been exterminated, Darwin noted, but the remaining Indians had become more barbarous. They no longer lived a traditional life in large villages, surviving on hunting and fishing. Now they wandered the open plains, without home or fixed occupation. For this reason, Darwin considered religious instruction by the missionaries and removing the inhabitants from their homelands as wholly positive.

Charles Lyell also argued that proximity to the natural world increased the likelihood of reversion to a prior savageness. While meditating on the domesticated dog, Lyell observed that dogs introduced to South America and then abandoned had gone wild, losing all marks of domesticity, reverting to the traits of their wilder originals. Yet he maintained that they did not truly become wolves. They were extremely savage, their ravages feared almost as much as those of wolves, but if their puppies were caught and raised away from the woods, they grew into obedient dogs.

Several millennia before, Aristotle had considered the re-emergence of the true nature of a thing when returned to its origins by imagining a wooden bed planted in the earth and springing up as a tree rather than another bed. Could immersion from birth in the wild cause the wilder nature of humanity to regrow? Darwin's theory of reversion turned out to be wrong; but the idea of it had infiltrated the imagination of thinkers in industrialized countries as the calamitous return to wilder nature.

The final stages of Jemmy Button's life seemed to confirm the tendency to revert to earlier savagery when reimmersed in the wildness of the unmanaged landscape. After leaving Keppel Island, Jemmy returned to the shores of Tierra del Fuego in 1858. The same year, George Packenham Despard persuaded nine Fuegians from Jemmy's tribe to journey to England. To all appearances, the group became painfully homesick and were eventually restored to Wulaia, a bay on Isla Navarino in Tierra del Fuego, in the autumn of the following year. Only a few days after the *Allen Gardiner* anchored at the bay so they could go ashore, several Fuegians emerged from their encampments and clubbed the missionaries and most of the schooner's crew to death. Jemmy was rumoured to have led the massacre, but when pressed at a hearing in 1860, he denied any involvement. If Jemmy was among those who murdered the missionaries and sailors, was it 'civilized' Jemmy or the native Orundellico who committed these acts? Despite his education in an Anglican school, his passable use of English, and the civilizing influence of his sojourn among European society, once he was reabsorbed into his native landscape in Tierra del Fuego, his original traits and knowledge reasserted themselves. Since the native people had not acted this way before, it was most likely that those who took part in the killings associated the schooner with abduction and that the attack was an attempt to prevent such an occurrence from happening again. But, for those in the industrialized world, Jemmy's life became a potent symbol of the benefits of a more tightly controlled environment.

After I arrived safely on the neighbouring island to Keppel, the storm broke. Sheets of rain seemed to lodge in the earth for a fraction of a second like thrown axes before melting into the ground. I peered out through the dappled, seal-coloured light. The island community amounted to no more than a few farms and a small, pretty stone cottage, with its cheerful yellow corrugated roof, glass porch and neat white picket fence. Two white socks left out on the clothes line kicked the air angrily as the strong winds swept across the treeless landscape. As everywhere in the Falkland Islands, Union Jacks were displayed as reminders of recently contested ground. One of the houses had white corrugated cladding, blue window frames and awnings, and a red corrugated roof, as if a larger flag overlay it. Even the outdoor privy was a Union Jack. Only the brick chimney kept the neutral tones of the earth.

I had travelled far from home in search of answers, and I often felt the dizzying longing for familiarity. Now, locked up by the rain inside the farm where I was staying, I stared morbidly at the obscured landscape, suddenly and painfully missing everything at home.

It saddened me to acknowledge that these feelings would doubtless evaporate on my return. There, I would find myself on a motorway, staring at embankments of stunted shrubs and nettles, empty packets of crisps and cigarettes thrown from car windows, at the taciturn regularity of intensive farmland, or the huge, featureless edifice of an out-of-town shopping centre. I would elbow my way on to a train alongside tired, grey-faced commuters, each with handfuls of packaged food, sandwiches, chocolate bars, bottles of water, which they would leave, half-consumed, in their places as they alighted. Even in the countryside, I would probably fail to escape the tinnitus of continual traffic on the thousands of miles of roads transfixing the landscape. Stealthily, a feeling of entrapment would wind around me, urging me to get away.

I liked the Falkland Islands because these settlements did not dominate the scene. The population was less than three thousand people, and the gorse and the penguins staked a stronger claim to the landscape. A comic sign on the door of a public convenience at Volunteer Point near Stanley showed a penguin standing on top of the toilet. The sign read, FALKLAND ISLANDS, QUIET, PEACEFUL AND SERENE. BUT BEWARE NO MATTER WHERE YOU GO, UP POPS ANOTHER PENGUIN!!!! Britain, by contrast, was in thrall to human needs, its wilder occupants harried from sight. Perhaps like those wanderers at the outer edge of an overcrowded territory, I began to dream of richer, untamed landscapes. It was almost impossible to encounter these in Britain without finding oneself engaged unwittingly in an adventure sport.

I knew that my interest in those whose cultures were once entwined with their native landscape reflected my anxiety about my own disengagement from place. I had visited one of the wildest places in the world. I had seen the incredible hostility of the landscape, experienced the laboriousness of trying to survive there. I had stared at the deadening particles of snow and ice, a world entirely inhospitable to us but to which an array of spectacular creatures had adapted. It was hard to know where I belonged. I wanted somewhere less cultivated than my home. I wanted, perhaps, what my home might once have been, or some scrambled blend of history and imagination.

Each of the individuals in the story of Keppel Island had experienced powerful nostalgia based on both real and imagined expectations of home. Forcing on Jemmy the wretched state of dislocation from place and people would have provoked strong urges to reunite with the landscape to which his memories and skills belonged. In contrast the Cornish fishermen died of a catastrophic failure to adapt their hunting skills to an unknown country. The man that led them, like so many zealots, saw no irony in their reversal of fortunes, each of the missionaries shivering, filthy and starving, as if the sea journey had hauled back the tides of time, leaving them in a state of nature, helpless against the elements. In a new landscape, relying on the knowledge of their prior homes hastened the deaths of the three fishermen from Paul. Despite being highly skilled in seafaring and fishing, they could not master the violent waters of Tierra del Fuego. They would have fared better to imitate the Fuegians, with their long traditions of knowledge of the region. Yet arrogance and blind faith in the superiority of their culture and worldview had fatal consequences. But then again, it would have been difficult to relinquish the habits of their home, for in the ancient past our ancestors had discovered that forgetting earlier habits could lead to extinction.

Darwin also became familiar with nostalgia on his voyage. He suffered the pangs of an unrelenting wish to return home that never diminished in all the long years of his voyage on the *Beagle*. 'I am sometimes afraid,' he wrote to his friend John Henslow on 18 May 1832, 'I shall never be able to hold out for the whole voyage.' Memories of his homeland haunted him. He began to fear that his new experiences would stifle his memories and he laboured in letters and diaries to keep them alive. This anxiety worsened as Darwin witnessed Jemmy's difficulties with homecoming. At other times, he was intoxicated by the dangers of being at sea, and in these moments, he prided himself on his ability to conjure up his old life in Cambridge vividly. In a letter to his sister Caroline, of September 1833, he wrote, 'I often think of the Garden at home as a paradise ... On a fine summer's evening, when the birds are singing, how I should enjoy to appear, like a Ghost amongst you, whilst working with the flowers.'

The Third Peregrination

North Yorkshire,
Manhattan Island
and Baffin Island

7

Bones

I arrived back in England in the spring of 2008. For the first
week after I returned, I searched for the edges of normality
and tried to peg them in place. I washed clothes still stiff with
the salt of the Southern Ocean. I went for walks along familiar paths, checking for changes in the old shapes and surfaces
of home. The otherworldliness of the voyage to Antarctica
continued to grip me as if a layer of its ancient ice shut me
off from my old life.

After my experiences on Keppel Island, I decided to look
for cultures that were vanishing nearer to home. It wasn't only
the traditions of indigenous societies like the Fuegians' that
were disappearing, but thousands of ways of life that captured
the idiosyncrasies of a particular place. I wanted to explore
the way in which variance was being ironed out in my own
landscape. In my research on whaling, the old port of Whitby
stood out in the narratives. Now a popular holiday resort
in the north of England, it was once at the heart of British
whaling. A few months after returning from Antarctica, I visited the town for the first time and was immediately struck
by the atmosphere. This was another landscape of extinction,
a region strangely captivating in its losses.

I spent the morning hours walking about the town. It was beginning to rain softly. Like a shell-seeker collecting in their palm a miniature of the beach, I picked up fragments of Whitby's past. Next to a narrow street called Arguments Yard was 'Harpooner Cottage', trimmed in blue wood with two portholes set into stone. There was a photographer who dressed clients in Victorian garb for sepia portraits. There was a shop called 'Natural Wonders', crammed with fossils, a few doors down from the Captain Cook Museum. Jet workshops specializing in contemporary jewellery haggled on one short stretch. I walked out of the town towards the piers that flanked the harbour mouth – two great limbs of yellow sandstone holding back the North Sea, which puffed in the distance. A plaque informed me that I was on the Haggerlythe cliffs, stabilized to prevent further erosion by the sea at the cost of several million pounds. On the opposite cliffs were amusement arcades garlanded in orange and pink. A yellow rescue helicopter streaked across the sky, scattering seabirds, as it disappeared out to sea.

In the afternoon, I visited Whitby library to search for old plans of the town. One of the earliest hand-drawn maps was a guide for mariners, dated 1740. A man's spidery black handwriting gave detailed directions for safe anchorage and for avoiding several of the hazardous rocks on a ship's approach. It inked in the confines of a tightly packed little seafaring village surrounded on three sides by almost impenetrable moorland. I sat at a small table in the back room of the library, comparing the maps as the sun impaled me through the windows.

John Ray, who had described the tin industry in Cornwall in 1674 in his *Collection of English Words Not Generally Used*, also told of the alum works in North Yorkshire. The extraction of alum from shale deposits triggered the growth of Whitby into one of the largest ports of the eighteenth century. Formerly it was a small, isolated fishing community where boats put to sea from wooden piers; the prosperity from alum led to the reconstruction of the harbour in stone so that industrialists could arrange deliveries of coal and trade their alum elsewhere. From there on, many of the buildings of the town were the homes of sailors, fishermen, merchants, or the sites of sailcloth manufacturers, chandleries and tanneries. In the eighteenth century William Scoresby drew a sketch map showing the proposed growth of Whitby. Under the tidal streams of the River Esk, he noted, 'To make room for ships, remove the High Bell (a dry area at half-ebb) & straighten the adjacent shores, and raise above the tide the Low Bell and erect a much wanted market place thereon.' Another map, less than a century later, showed several shipyards and docks, the roperies, the tannery, and the occupied homes and businesses clustered together, although little had significantly changed but for the construction of the Whitby and Pickering Railway. By the time of the first Ordnance Survey map of 1895, most of the shipyards were gone, replaced by buildings and infrastructure of a more general nature, along with housing estates and a school. The road beside the defunct ropery had been renamed Rope Walk, and the railway had expanded to include a general goods yard. There was a spa and tennis courts on the front. The first technically drawn map, dating from 1948, showed the considerable changes that had taken place in only fifty years. Much of the infrastructure of the maritime industries had disappeared, replaced by leisure facilities – the Waterloo Cinema, a band stand, a miniature golf course, and shops selling an array of products. The distinctive way of life determined by the natural situation of the town amid sea and shale had largely vanished; hotels and guesthouses were constructed side by side on the West Cliff and the sailors' and fishermen's cottages turned into holiday homes with names like 'Sea Shanty', 'Crows Nest' and 'Anchorage'.

From early on, fishermen and merchants colonized coastal places, where the gaze as a matter of course looked beyond the range of home. There was no incentive to seek out the littoral but for trade. All that changed around 1750, when a Dr Russel, the son of a London bookseller, wrote about the beneficial qualities of sea water, especially for glandular complaints. Almost immediately, people sought out British beaches 'like so many land-crabs'. Dr Russel's patients flocked to him in Brighton, and old communities, whose fortunes were failing, became prosperous again. Seaside towns were reinvented as salt-water spas – places of amusement and recuperation for the increasingly affluent and mobile masses.

The following day, I took a walk beside the old railway track that ran alongside the Esk estuary. The hedges were overgrown, their branches crucified by the wire-mesh fence. I walked silently along the muddy path as the sounds of the town softened to a low hum. I passed a small boatyard, where rows of sailing and fishing boats rested on trailers or wooden cradles. Until the latter half of the nineteenth century, local ship carpenters and sailors worked from dawn until dusk, called to and released from work by the chiming of the Town Hall bell, singing shanties to ensure the hauling out ran in good time and rhythm. I had read in Whitby library that Jarvis Coates operated the earliest shipyard in Whitby at a place formerly called Walker Sands, bordering on Bagdale Beck. Coates died in 1739, succeeded by his son Benjamin. His family eventually sold the business and the last registered owner was a Mr Barrick, who gave up the yard when he went blind. After its abandonment, the North-eastern Railway Company purchased the land in 1865, filled the dock and obliterated all signs of its former life as a shipyard. Jarvis Coates's eldest son, also named Jarvis, set up the second shipyard in Whitby. In 1836, the office of this shipyard became that of the Whitby and Pickering Railway Company, and, within a few years, nearly every trace of this shipyard had also been eradicated.

Further along the path were old wooden sheds from which smaller boats were built or repaired. From one of them emerged a man dressed in blue overalls who invited me in. Dave had worked with boats in one way or another for much of his life, and he was now transforming the shed for his retirement. The shadowy, dust-filled downstairs was framed like a boat workshop. I recognized the smell of heated metal whetting wood from my father's shed, and its sharp scent suffused the place with familiarity. Up a set of freshly carpentered wooden stairs was a large, sunlit room. Dave had constructed the room masterfully, almost like the inside of a boat, using old portholes to bring extra light through the back wall. At the front was a small balcony that looked out on to the River Esk and a garden off to one side below. I envied Dave's boathouse – its ingrained spareness and coherence, the soothing and organic beauty of its view over the river and out to the sea beyond. It expressed his interests and his occupation in a homely structure that wanted for little else.

We flicked through a scrapbook with photos and newspaper cuttings about his former sailing coble, *Gratitude*. She was fitted with a traditional lug sail, and for a while she was one of the only classic cobles left. According to Thomas Hinderwell's history of the nearby town of Scarborough, written in 1798, cobles, unique to the north-east coast of England and the borders of Scotland, landed here with the Vikings. Formed of oak and larch, cobles were shaped to launch from a coastline with few natural harbours. As the gradient of each beach and the swirl of the tide around the rocks were so individual, local builders most often adapted the basic design to meet the special demands of the fisherman and his particular fishing grounds.

In the Scarborough Maritime Museum, a letter from the early seventeenth century to Sir Thomas Chaloner of the Yorkshire town of Guisborough describes the distinctiveness of cobles and the perils of landing the boats on these shores. As the fishermen were 'acquainted with these seas', on seeing a huge wave ready to overcome them, they would 'suddenly oppose the prowe or sharpe ende of theyre boate unto yt, and mountinge to the top, descende downe as yt were unto a valley. Whereupon mountinge with their cobble as it were a great furious horse, they rowe with might and mayne, and together with that wave drive themselves on lande.'

The boats' distinctive square stern and sharp prow made them good for rowing as well as sailing. Fishermen beached cobles stern-to, so an oblique transom was less obstructive, and, without a keel, they were easier to handle on shore. Dialect words catalogued with precision the complexities of the coble's design. The 'scut' of the stern, the 'knee' of its supporting timber. 'Kilp' denoted the curve of the rear underside; 'gripe', the depth of the coble at the forefoot; while the 'tumblehome' was the inward curve of the upper part of the boat.

Commonly, three fishermen went to sea in one coble, each with three lines spun around a wicker frame, and baited hooks fastened with twisted horsehair. These long-lines were shot across the current. In the eighteenth century, Hinderwell remarked that the flood and ebb of the tide was so rapid that fishermen could do little else but leave the lines and wait. Two of them wrapped themselves in the sail and slept while the other kept a vigilant watch for swell.

In 2004 one Whitby fisherman, Shaun Elwick, still went long-lining during the winter from his coble, *Carisma*, fishing for codling. Although his coble was motorized, it remained an unpredictable endeavour, he explained, as the boat trialled against the sea and the tides. In 2008 a film by Sarah Macmillan portrayed a man from Redcar, a fishing town in

Cleveland, just up the coast from North Yorkshire, hewing models of cobles from oak. 'Nowadays there isn't many cobles left,' he said, 'because it isn't a paying game and cobles don't live for ever.' Kelvin engines were fitted from the 1930s onwards, and boat builders modernized the coble's design, adding covers to protect the engines and sonar, but the coble was essentially formed for sailing and rowing. With an engine, its special shape was no longer needed. The old cobles depended on an almost fraternal closeness between the sailor and the seas. Without that intimate knowledge of the response of the boat to the elements, the risks of setting out in a coble on the undependable North Sea would be much too great. As knowledge of the seas and weather declined, and the wages

and fish stocks diminished, traditional cobles like the *Grati-tude* steadily disappeared from the coastline.

Cobles were small boats suited to a team of a few fisher-men. When larger, more effective vessels like doggers, robust Dutch boats trawling nets behind them, began operating in the North Sea, the region's commercial fishing surged. Boats of far greater size and capacity needed outside investment, financed by larger companies. Coble fishermen worked sea-sonally, in summer trying for herring, which migrated down the east coast of Britain, in winter crabbing or long-lining for cod. The involvement of investors with stakes in larger boats was based on targets and profits that would cover their out-lay rather than on sustainable or realistic returns. Companies called for a year-round business rather than the seasonal en-deavours of smaller operations. Large, commercial ships went to sea for long stints, freezing their catch on board. By the 1960s, the North Sea herring stocks were plummeting. And, by 1977, a moratorium on the herring fishery was brought into force. Angry fishermen pinned the blame for its collapse on the failure of the North Sea countries to cooperate so that their fishing fleets could catch a sustainable harvest. It was an all too familiar story.

I waved goodbye to Dave and carried on along the path to the old gasworks. I was trying to find the site of an old canvas shed I had seen in a smoggy photograph from the 1930s in the archives of the local paper, the *Whitby Gazette*. The paper gave its rough location, somewhere towards the Whitby Gas Company's works, close to the railway line and the viaduct.

Dave had told me, 'That shed used to be in Johnno's place, but there's no sign of it now. But go and ask him, and I'm sure he'll let you have a look around.'

The reason for my curiosity was that the shed was built from the huge jaws of a whale. Throughout England and Scotland people once used whale bones as cruck beams, or for bed frames, for chairs and tables, for window gratings. In his old age, Darwin even leant on a whale-bone walking stick, his hand clasped around a pommel shaped like a skull with eyes of emerald-coloured glass. Whalebone from baleen, the filtering structure inside the mouths of some whale species, once occupied the material niche now filled by a multitude of synthetic plastics. A 'singular substance' that when heated retained the shape given to it, it stiffened corsets, the brims of hats, caps and bonnets, the bottoms of sieves and riddles, and lined the backs of sofas. Whaling, as practised by British seamen, was never a subsistence activity. From its inception, it was purely driven by commercial interest. A Scottish whaling company received the first official licence to whale from Charles I in 1626, to compete with the Dutch and Spanish companies that hunted bowhead whales close to the shores of Spitsbergen. In the middle of the eighteenth century, the industry pushed further west to new hunting grounds in the Arctic. By this time, the oil rendered from blubber was in huge demand as the main fuel for street lighting in Europe's major cities and as a lubricant in the manufacture of machinery.

The difficulties of harvesting whales in sufficient quantities

to meet demand required backers willing to put up money in the expectation of favourable returns. The cost of the ship's hull alone was sizeable. After this, whaling companies had to pay for sails and masts, riggers, cordage, ironwork and timber, plumbers and glaziers, fuel and provisions, along with insurance and the salaries of the crew.

The first whalers to enter the trade from Whitby were the *Sea Nymph* and the *Henry and Mary*, which sailed for Greenland in 1753. Until 1837, around twenty-five thousand seals, over fifty polar bears, and just under three thousand whales were caught, processed and shipped back to the town. In those days, four large oil houses stood alongside the Esk, where the blubber brought back in casks was boiled. The whalers would attach bones to the front of the ship if the boat had a successful catch and had lost no crew, and, in a matter of days, the stench of boiling whale blubber would suffocate the town. When the blubber was boiled in the Arctic, wrote William Scoresby the younger, the oil was brighter, paler and more inflammable. It was sweeter smelling than the putrescent gases burning in British towns.

Scoresby's father, another William, inventor of the crow's nest and a primitive ice drill, was born in 1760 and worked as a farmhand until he was apprenticed in his late teens to the Whitby ship *Jane*. In 1787, he sailed from the town on his first whaling voyage to Greenland in the *Henrietta*, commanded by Captain Crispin Bean. Only four years later, he was captaining the whaler himself. His first voyage was a failure. Of the seven ships that left Whitby that season, one was lost, two returned with only one whale, 'one of them very small', and four of them returned 'clean', with nothing. But the following year, Scoresby arrived home with eighteen whales, a yield of over a hundred tons of oil. His son followed him into the trade, and left an account of the whaling industry in his journals written on a voyage to the Arctic in 1820. He described the steady retreat out to sea of the bowhead whales, from the shores around Spitsbergen to the banks and then to the ice edge. Under pressure from whalers, the bowhead population then migrated further, to sanctuaries shielded by the ice. 'In consequence of this event,' Scoresby wrote, 'the plan of prosecuting the fishery . . . now underwent a material change. This change did not only affect the manner of conducting the fishery, but it likewise extended to the construction of the

APPARATUS used in the WHALE FISHERY. PLATE

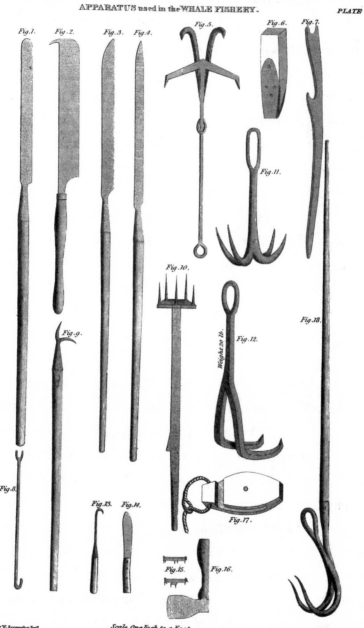

Fig.1. Fig.2. Fig.3. Fig.4. Fig.5. Fig.6. Fig.7.

Fig.11.

Fig.10.

Weight 20 lb. Fig.12.

Fig.9.

Fig.18.

Fig.8.

Fig.13. Fig.14.

Fig.17.

Fig.15. Fig.16.

Drawn by W. Scoresby Jun.ͬ Scale. One Inch to a Foot. W.&D. Lizars Sculp.ͭ

ships, and the quality and quantity of the fishing apparatus.'

The younger Scoresby was a colourful character with an insatiable interest in the natural sciences. As a whaler, he studied the meteorological conditions in the Arctic, making detailed observations of snow and crystal formations. He wrote to Joseph Banks about his discovery that the depths of the polar ocean were warmer than the surface temperature. Despite becoming ordained as a minister, he contributed over sixty papers to the Royal Society, and sailed to Australia in his sixties in a bid to settle a dispute between him and Sir George Bidell Airy, the Astronomer Royal, concerning the means of correcting the effects on the ship's compass of the iron used to build ships. Scoresby died of a massive heart attack on return from the arduous voyage. One of his many legacies was the Greenland Magnet to assist the tricky and highly dangerous navigation through the ice.

Most years, some of the ships failed to return, or returned home with the difficult news that members of the crew had perished. In 1820, eleven ships from Whitby and sixty-two from Hull, another historical whaling port further down the coast, found themselves hemmed in by the ice in Baffin Bay. Amid the traumatizing restrictions of the polar ice, a peculiar whaling culture sprang up. The frightened sailors invented many things, from more effective harpoons to stirrups that the flensers donned to walk across the whale carcass. As a pastime, they carved whale teeth and bones with images of winsome women, whales and rigged ships, some rough-hewn and racy, others highly ornate. Known as 'scrimshaw', the art form came to have its own commercial value. It was probably inspired by the traditional carvings of the Inuit people of the Arctic.

Whitby's whaling industry went from boom to bust in a comparatively short space of time, foundering as early as 1837. Although whalers from the town feared that their livelihoods would become unprofitable as gas from the distillation of coal replaced the oil or oil-gas from whale blubber, it was the rapidly diminishing stocks of whales that truly weakened the industry to the point of collapse. The whale-bone structures like the shed I saw in the old photograph were remnants of an era of plenty. By the late nineteenth century, cheap fossil fuels combined with the collapsing whale stocks brought the entire northern whaling industry to its knees. By the first decades of the twentieth century, the bowhead whale was almost extinct. Collectors began to snatch up scrimshaw for museums and private collections. By the 1960s, fake scrimshaws carved in plastic were circulating, as the demand for the originals outstripped supplies. By the early part of the twenty-first century, whale-watching tourism was more lucrative than the original industry.

For the people of Whitby, the residues of earlier ways of life were reduced to the curios of tourism or to the leisure industry. Much that was uncommon about the place was gone or forced to the margins. The whale-bone shed once stood on a flank of land that marked a people's relationship to the sea. Now, for most of the inhabitants, such a relationship with nature and the elements was absent, and many of the new structures built atop this past could be seen in any town in the developed world.

While the inhabitants had knowledge of the natural environment, the buildings and tools that surrounded local people prompted a regular recognition of their closeness to nature. But this relationship changed over the centuries from one of survival and subsistence to profit. Each activity foundered in its turn, first with the alum quarrying and later the maritime industries. The physical and cultural memory of this earlier interdependence with nature began to fade. Some buildings changed use almost immediately, utilized by associated trades that still prospered, such as the whale blubber cookery that became a smokehouse for preserving herring as kippers. Others steadily retreated, buried beneath new structures and aspirations. The many whale-bone arches that framed a kind of prayer for reminiscence across Whitby were dismantled bit by bit or else crumbled to dust in the excoriating sea air. In the 1960s, a single arch was set on the West Cliff, with a plaque that stated, LITTLE IS NOW LEFT TO REMIND US OF THESE PROSPEROUS DAYS OTHER THAN THIS PAIR OF WHALE JAWBONES.

Having found no sign of the whale-bone arch in Johnno's garden, I then went in search of one of the old gasometers. I found a workshop, where a group of men in smudged overalls were repairing cars. Behind one vehicle, elevated so that the worker could slide underneath, was the large cylindrical space once used to store gas, a smoke-grey curvature of stone, its cracks invaded by mosses and weeds. Neglected and monopolized by other priorities, this ruin from Whitby's whaling history indicated only forgetfulness.

Several days later, I was on my way to the famous ruins of Whitby Abbey. I had eaten a warming lunch in a café on Church Street called Humble Pie 'n' Mash. The owners prided themselves on traditional home-baking and told me that, as far as they knew, in the past the building had been a cobbler's, a dressmaker's and a bakery. A glass cabinet displayed rusted buckles, scissors, hairpins: trinkets salvaged from these erstwhile activities. The décor was carefully reminiscent of the 1950s – blue-checked cotton tablecloths, simple, straight-backed wooden chairs carved with hearts – invoking an earlier simplicity to which people now harked back.

I passed some of the town's old workshops, since converted into a string of orderly, red-brick family homes, the outlines of television sets, china ornaments and the guileless faces of the knowingly alone just visible in the darkened interiors. As I ascended Caedmon's Trod, the footpath leading to the headland, I could see the harbour of Whitby below. A light sea mist scribbled out sections of the landscape as it drifted, but the scene wasn't melancholy or fantastical. Whitby Abbey was ahead of me, its magnificent ruins altering suddenly in the sunlight from austere pewter to a maze of corals. In the seventh century, it was the seat of the Northumbrian nun Hilda, who rose to such eminence that this was where the synod met when representatives of the Celtic and Roman strands of British Christianity decided on a common date for

Easter, a decision that many believe led to the unification not only of two ecclesiastical factions but of England as a whole. After invading Vikings wrecked the first abbey in the ninth century, a second was built from around 1220, falling to ruins, too, in due course, neglected after Henry VIII's dissolution of the monasteries, ravaged by storms, and by German naval shelling in 1914. The abbey still dominated the cliffs, an ideal navigational aid, its stone edifice jinxed by the moon.

In Bram Stoker's *Dracula*, the archetypal vampire story partly written and set in Whitby, the ruins of the abbey epitomized the broken-down defences of England against the intrusion of more savage elements. My father handed me a copy of the book when I was about thirteen or fourteen, telling me that it had frightened him more than any other book he'd read as a boy. I could have no greater encouragement to read it, and it became one of my favourite childhood books. In the novel, a newspaper cutting about Whitby tells the reader of curious happenings during a sudden and

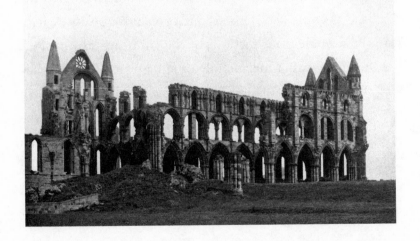

mighty storm. A ship in distress on the waves becomes visible as a cloudy silhouette. When it nears the shore, the onlookers catch sight of a corpse lashed to the helm, which turns out to be the captain of the vessel. The ship is devoid of life, except for what appears to be a large dog that springs from the deck on to the shore, melting into the storm's gloom.

This story had something in its dark myth that connected it to the extinction of cultures. One of the chief anxieties about Dracula expressed by Jonathan Harker, the novel's hero, who encounters the monster first on his home turf, is the skill with which the count assumes the behaviour and cultural mores of the English. Dracula is concerned to subdue his origins and attain an air of Englishness in order to infiltrate himself into society, and his appearance is masterly. None the less, he can never abolish his true, bestial nature. By the eighteenth century, European thinkers like Johann Herder, the French naturalist de Buffon and the German physician Johann Blumenbach generally agreed that savagery in other societies was the result of migrations that forced some people to live in inhospitable regions over whose natural elements they had little control. However, these thinkers assumed that all people, deriving as they did from God's design, could rise to a state of civilization by escaping the wretched conditions of their environment. But the nightmarish visions of human mongrels in *Dracula* reflected a shift in thought, an emphasis on intrinsic biological differences that explained the divergence of societies.

Dissections carried out by anatomists during the previous

century revealed the hemispheres of the brain, the shocking, gunmetal border that split the two masses of the organ of thought. Franz Joseph Gall established the study of phrenology, which allocated multiple attributes and functions to different parts of the brain. Earlier, in the eighteenth century, the Dutch anatomist Pieter Camper had suggested that the heads of animals gave insight into their biological nature, while the British naturalist Joseph Banks had become enthralled by the anatomical differences between aboriginal peoples. He encountered a number of tribes and societies as he led the scientific group that accompanied James Cook's first world voyage aboard the *Endeavour*. On his return to England, Banks was involved in the grisly business of acquiring aboriginal skulls, most especially those of native Australians. More often than not, these skulls were obtained through violent skirmishes between indigenous and colonial parties, and in some cases through outright murder.

In an early nineteenth-century review of the new field of phrenology, Francis Jeffrey summarized George Combe's proposition that 'the degree in which any man possesses any intellectual faculty – moral virtue, vice, or propensity – nay, any animal emotion or power of external sense or perception, or even, as we take it, any acquired habit, infirmity, or accomplishment – may be certainly known by the size of certain protuberances on his skull.' Phrenologists maintained that these protuberances corresponded to separate regions of the brain, whose size determined the strength of certain habits or propensities. Phrenology, combined with the theories of anatomists, fostered the belief that the external differences among peoples reflected the quality of their innate features. Where before an indigenous person shared a seed of civilization that removal from the wild environment and sufficient education might germinate, now different races were viewed as biologically stuck, physiologically hindered from rising to the heights of supposedly superior humans.

The obvious contrasts between European skulls and those of the aboriginal Australians, from thicker foreheads to large nasal cavities, fuelled a new prejudice that such differences were indicative of the lesser intellectual strength and capabilities of other races. Such people were by nature incapable of any rise in status. While some commentators at the time recognized that imported diseases and vices from Europe, along with violent persecution, were contributing to the decline of some indigenous societies, others became convinced that these people were doomed to extinction by dint of their inferiority.

For contemporary readers of *Dracula*, the story derived its power from fears, perhaps especially prevalent in Britain as unrest in its colonies mounted, that some dangerous and crude vigorousness in different nations and cultures threatened their sophisticated life. Bram Stoker wrote his novel before the start of the First World War, when the Transylvania from which Dracula originated was still a part of the Austro-Hungarian Empire. By locating his monster in Transylvania, Stoker was alluding to the so-called 'Eastern Question', which preoccupied European politics throughout the

nineteenth century and remained unresolved until the collapse of the empire at the close of the First World War. The
gradual disintegration of the Ottoman Empire led to violent
unrest in the Near East and shook Britain's belief in the sta-
bility of its own imperial territories. The expansion of the
British Empire was seen in the ideology of the time as an
experiment in civilizing other peoples. Faced with growing insurrection among indigenous populations who were
tired of adventitious governance, colonial powers like Britain
interpreted their behaviour as the resurgence of inherently
lawless characteristics.

Contemporary reports of struggles for control in parts of
the Near East emphasized the racial differences of the victims and adversaries, and the savage nature of the crimes. In
a book published in 1896, after thousands of Armenians re-
belling against Ottoman rule had been massacred in 1894
and 1896, the Reverend Edwin Bliss wrote of the Armenians as once numbering around three million people. 'In the
turmoil of the centuries, they had been scattered until their
ancestral valleys and mountain slopes have largely passed into
other hands. They still preserve, however, the racial characteristics of that early time, and look back with intense yearning to that olden time and those familiar places.' Bliss then
wrote of the 'rich and powerful potentate of another race'
that now controlled their lives; the Turkish rulers, he claimed,
showed abominable cruelty in reaction to the resistance of
the Armenians.

One of Bram Stoker's characters, Mina Murray, noted of

her friend Lucy Westenra that some individuals had 'too super-sensitive a nature to go through the world without trouble.' Transfusions of human blood were necessary to guard against any reversion of the super-sensitive to a savage or susceptible state. In his 1909 *Essays on Eugenics*, Francis Galton wrote that 'The possibility of improving the race of a nation depends on the power of increasing the productivity of the best stock.' He had coined the word 'eugenic' in 1883 to mean 'having good inherited characteristics', convinced that encouraging the 'best' people to breed would eradicate potential animal traits. 'The general tone of domestic, social and political life would be higher. The race as a whole would be less foolish, less frivolous, less excitable and potentially more provident than now.'

But Dracula's convincing adoption of English mores raised the question of whether so-called 'civilized' societies were capable of cloaking monstrous possibilities beneath urbanity's garb. If human brains were a patchwork of animal and cultured impulses, of new gifts and neglected memories, how might any person either appraise or control their behaviour? Under what conditions could any individual or society abandon its advances, returning to cruder or wilder manners?

The Armenian massacres that the newspapers in Europe and America described in such stirring detail served as evidence of the depravities of which all people were capable. At the time, the Muslim leaders of the Ottoman Empire justified their suppression of a people by citing their different religious beliefs. But the onset of the First World War led to the extermination of the Armenian people of Anatolia in death camps and marches, especially in the deserts of Syria. The organization of these mass killings was facilitated by the advent of the telegraph and railroads, which allowed a greater degree of coordination.

The pendulum swung heavily towards belief in control, civilization and education as the means to peace and prosperity. Earlier thinkers came back into favour. In 1874 John Stuart Mill had stated that 'it remains true that nearly every respectable attribute of humanity is the result not of instinct, but of a victory over instinct'. He believed that anything worthy in human nature derived from a kind of taming. In his digression on the term 'natural', Mill postulated that since the natural elements and wild beasts perpetrated horrors, people should not imitate the model of nature. 'Either it is right that we should kill because nature kills; torture because nature tortures; ruin and devastate because nature does the like; or we ought not to consider at all what nature does, but what it is good to do.' Mill maintained that the goodness of our actions could not emerge without stringent separation from the natural world and the rustic, unrefined lifestyles associated with it.

I paused briefly at the top of the 199 steps that descended from the headland to the old town of Whitby, as a little girl shakily took a photograph of her parents. Turning to me, she said, 'Thank you,' with the formality of the very young. Dozens of tourists were wandering through the cobbled streets below, browsing in shops and eating ice cream despite the greyness of the afternoon. I slipped into the crowd that meandered down towards the river, looking wistfully at the jars of assorted pear-drops, chocolate limes, humbugs and liquorice that lined the walls of a traditional sweet shop. Whitby was peddling nostalgia and yet there was something genuinely suggestive in its commercialization of the past.

I walked across the bridge, looking out beyond the piers to the smoky upheaval of the North Sea. The river divided two quite different stretches of the town. To the east, the narrow streets and old buildings were self-consciously reminiscent of earlier decades, while the west was much like most of the towns of Britain, with several large supermarkets, Asian take-away restaurants, pizza places, gyms, offices, banks and travel agents. Along the harbour front, there were fish and chip shops and the 'Dracula Experience Tourist Attraction', and the side alleys were full of souvenir shops with retro or maritime themes like 'The Seafarer' or 'Nautilus Gifts', or tattooists and shops full of black lace corsets and floor-length coats, the paraphernalia of the Gothic subculture that proliferated around Whitby as a result of *Dracula*.

The Goths had thrown a banderole of rebellion across the town. Although itself an enormously popular and international phenomenon, arising as it did from the punk movement of the 1970s and borrowing promiscuously from history for its tokens and regalia, the subculture of the whey-faced Goths seemed to me an important resistance to the escalating homogeneity of many modern lives and landscapes. Their costumes and outlook celebrated the freakish version of human nature conjured up by Gothic writers and film-makers.

After the radical social transformations of the First World War, the metaphors of nineteenth-century Gothic novels infiltrated early experiments with film, most especially the horror films of German expressionist cinema, such as *The Cabinet of Dr Caligari*. A silent film from 1920, its fuzzy black prologue announced that the story was 'a tale of the modern re-appearance of an 11th century myth involving the strange and mysterious influence of a mountebank monk over a somnambulist'. The film's early scenes are of a travelling fair in which a traitorous-looking impresario, Dr Caligari, flamboyantly wakes a sleepwalker, Cesare, for the first time in twenty-five years, with the words, 'Wake up, Cesare. I, your master, command you!' Later that night, a series of mysterious murders takes place. One of the chief characters mouths, with the gravity of his silence, 'There is something frightful in our midst.' We, like the film's protagonists, are led to believe that Caligari is somehow steering Cesare to commit the murders as he sleepwalks the streets at night. Only in the closing scene do we discover that Caligari is in fact a doctor in charge of a mental asylum in which Cesare and the other protagonists are confined in the senselessness of their own insanity.

There is a related scene in *Dracula*. As the plot progresses, the abbey and churchyard on the cliff entice the black dog and Lucy Westenra, Dracula's first victim. When the beast attacks Lucy, she herself begins to transform at an intrinsic level into something fierce and unruly. After she leaves Whitby, her condition worsens. She alternates between a state of livid animation and one of anaemic lethargy. But when the doctor observing her draws and analyses some of her blood, he finds her in vigorous health. It must be something mental, he decides, as Lucy complains of troubling dreams that she forgets on waking.

In 1897, the year that *Dracula* was published, Sigmund Freud completed *The Interpretation of Dreams*, which first appeared in German in 1899. In the book, he wrote of early experiences hidden to the conscious memory re-emerging in dreams. For Freud, images of real places, persons and objects, if lost long ago, entered the sleeping mind as utterly bizarre and unfamiliar, until the revelation of their historical or physical origin. Taking himself as his example, he admitted, 'I was haunted by the picture of a very simply formed church tower which I could not recall having seen. I then suddenly recognized it with absolute certainty at a small station between Salzburg and Reichenhall.' He then described his recurring dream of standing in a singular place, sensible of some inexplicable darkness to his left from which grotesque sandstone figures loom. A decade later, he established the real source of this dream in Padua, when he happened to visit the city again and stumbled on the haunt. Such dreams, Freud concluded, drew on experiences from childhood forgotten in the adult's waking life.

The idea that the forgotten playgrounds of childhood were the source of recurrent or baffling dreams and feelings grew out of the sentiments of earlier writers like Godwin and Ferguson, who emphasized the closeness of our juvenile years to the early animal history of our species. At the turn of the twentieth century, an individual's dreams were seen as a throwback to the unknown and sometimes undesirable reaches of our animal past. As the habitual experience of natural settings became memory only, mere echoes and evocations that time or fresh aspirations could distort, this ambience of loss penetrated people's sleeping or idling minds, giving rise to wild dreams and misshapen realities.

The nineteenth-century writers who inspired film-makers between the wars had let loose macabre and primitive spectres of nature by exploiting the pejorative sense of the term 'gothic' as threatening, uncultured and barbarous. In medieval societies the new style of architecture was described as 'gothic' because it was seen as crude compared to classical buildings. But the nineteenth-century romantic revival coveted its atmospheric promise and distinctiveness. The barbarous overtones of the original term became muddled with an excited, inquisitive nostalgic sentiment.

In Whitby, as in thousands of places around the world, customs and conventions governed by the sea and earth had disappeared. In place of these extinctions was a less distinctive, man-made landscape, designed for ways of life less conspicuously reliant on nature. Karl Marx had already prophesied such a world, in which people ended up working, earning and purchasing to sustain their way of life, not true liberty but the eventual servitude to economic forces. He and others concerned by the rise of capitalism could do little to slow the process once it was underway. Many societies were busy writing over the influence of nature with new buildings and infrastructure that encouraged economic growth to govern inhabitants.

In the 1960s, phrases like 'the society of the spectacle' were coined to account for the stupor engendered by rapidly growing economies that suppressed the more natural and authentic needs of people's lives and instead cast a spell of uniformity over individual habits. Industries perpetually sought new means of acquiring profit for expansion, especially by installing their production lines overseas in poorer countries, then transporting their goods for purchase all around the Earth. The USA was the first nation fully to exploit the commercial potential of mass culture in the post-war years through the so-called Information Agency. An offshoot of the American government, the agency declared in its mission statement that its aim was to 'influence foreign publics in promotion of the national interest', which involved increasing familiarity with American products abroad.

In Whitby the shops on the west side of the harbour epitomized the trend towards globalization and uniformity, selling Hewlett-Packard computers, Del Monte canned foods, Exxon Mobil oil, Reebok trainers, Microsoft games and other ubiquitous international brands. To those who resisted such changes, these products represented the sheer avarice and materialism of the modern world. Yet it was not solely greed for money that had put economic forces at the centre of human lives. It was understandable that there was a growing movement away from nature when the world had seen the damage that man's more brutal instincts, magnified by modern technology, had meted out. People chose to uphold the view that only the modification of nature might bring some measure of order and happiness to human lives.

8

Tundra

In Whitby closeness to nature had been replaced by a proficient, implacable commercial culture. Following my time there, I sought out somewhere at an earlier point of alienation, a country or culture with the opportunity to choose an alternative course. The history of one such society, the Inuit of the Arctic, was tied to the Whitby whalers by a knot of mutual interest. Were the native traditions of the Inuit also on the threshold of disappearance, their society tugging away from the natural world as so many other cultures before them? I decided to meet these people for myself and see what remained of their knowledge of the Arctic ice and tundra.

In 1999, news broadcasts showed the jubilant celebrations of the Inuit of Canada when they achieved self-governance with the institution of the Nunavut territory. In the following years, reports became more frequent that the small, isolated Inuit settlements were gradually losing inhabitants to larger communities, and that the unfamiliar demands of wage employment and mass consumerism promoted by the global media had eroded something of their identity. Despite their new prosperity and autonomy, Inuit nations had disproportionately high levels of alcohol abuse and domestic violence, and a suicide rate four times that of the rest of Canada.

'Will the Inuit disappear from the face of the earth?' the past president of the Inuit Tapirisat of Canada, John Amagoalik, asked. 'Will our culture, language, and our attachment to nature be remembered only in history books? To realize that our people can be classified as an endangered species is very disturbing . . . If we are to survive as a race, we must have the understanding and patience of the dominant cultures of this country . . . We must teach our children their mother tongue. We must teach them the values which have guided our society over thousands of years.'

Before commercial whaling decimated bowhead whale populations and the industry collapsed, some ten thousand men had sailed to the Arctic from Europe and North America. After spending time in Whitby, I was curious to know how the influence of the whalers had infiltrated the Arctic landscape and its people, affecting their native awareness. In over fifty communities, Inuit still hunted several species of whale, along with caribou, walrus, seal, fox and ptarmigan, as they had done for many generations. Unlike commercial whalers, the Inuit traditionally regarded their annual harvests as a means of communal survival.

Many Inuit knew long before climate-change scientists that their landscape was altering. Hunters travelling across the sea ice were responsive to the dangers of ice conditions. Inuit hunters habitually predicted the weather by combining the impressions of their ancestors with their own experience of the landscape. Ancestral knowledge of successful adaptations to unfamiliar conditions encouraged adjustments to hunting and lowered the likelihood of failure. For decades, hunters spoke among themselves of the slow and tardy freezing of the sea ice, the untimely soup of its break-up, of diaphanous islands of ice and unexpected streaks of open water. The wind, too, had become skittish and unreliable. In the late twentieth century, scientists began turning to Inuit hunters, combining modern technologies with the unrivalled knowledge of those living from the ice. The anecdotal evidence of the hunters fused with satellite systems and mapping methods to give greater insight into the increasing fickleness of ice conditions across the Arctic.

Over the past eight thousand years, the Arctic received less summer-time energy from the sun, due to the slow wobble of the Earth on its axis. This cooling trend would ordinarily take another four thousand years to reverse. But records of the climate of Baffin Island have revealed that warming has already begun. In the summer of 2007, the extent of Arctic sea ice declined to unprecedented lows. Compared to the extent of the ice in 2005, the area lost was comparable to the states of Texas and California combined. At the same time, it was shown that the oldest and thickest sea ice in the Arctic was shrinking. The high Arctic, which ought to have stayed a frozen province for millennia, was threatened by becoming accessible to those intent on extracting its stores of oil and gas, constructing large amounts of infrastructure across the unspoiled tundra, ingraining pipelines into the rock.

Over several months, I planned my voyage from

Southampton to New York and onwards to Baffin Island in
Nunavut. I set off in the winter of 2009.

The ship drew into South Ferry Terminal in New York in the
early hours of the morning, the sky brindled by the jazzy lights
of buildings and advertisements. I had booked the train from
Penn Station in New York to Montreal in Canada. From there,
I planned to travel by a small turboprop plane into the northern territories of the Inuit. The Statue of Liberty was a mint
haze on the horizon, the colour of a faded dollar bill. Hundreds of brightly painted shipping containers paved the shoreline. Beyond these were giant stands of glass and metal, each
breeding human life in its reflections. At a distance, there was
no sign of greenery or wildlife except for a few birds.

Once in Manhattan, I turned at the corner of St Mark's
Place in the direction of East Village, passing Japanese, Mo-
roccan and French restaurants all on the same stretch, as if the
Earth's continental plates had shrunk to jigsaw pieces. Across
New York there were thousands of eateries trading in the
spice and fragrance of distant lives and countries. Culture and
identity had become big business. But I loved the city, the
infectious chaos of everything man-made and short-lived.
It was tempting just to throw myself into this bedlam of
chances and treats, and to take unadulterated pleasure in it.

In a letter to his sister Minette written in 1664, King
Charles II boasted of the rise in English subjects in America:
'You will have heard of our takeing New Amsterdame, which
lies just by New England. 'Tis a place of great importance

to trade, and a very good towne.' Prior to Charles's reign, in October 1606, King James I of England issued two charters allocating land to the Virginia Companies of Plymouth and London, a coastline stretching from Virginia to New York. Profit was the sole intention of these private ventures, which sought to plant the English flag and trade commodities. The first colony, James Town, was established in 1607 at the mouth of the Hudson river, close to present-day New York. Between the late fifteenth century and the middle of the sixteenth, England's population had more than doubled but the growth was even more conspicuous in London. 'People swarmed the land,' one bystander said, 'as young bees in a hive in June.' The solution was to transplant something of England in new, broader soils. By no means for the first time in history, a dominant power experimented with expanding beyond its original boundaries.

As I made my way through North America, I began to brood on the standard, statutory version of English that spread out from James I's era. It was a language that dulled sparkling local varieties, facilitating rule across a multitude from a single source of authority. It found different currency in the motto for the United States as *E Pluribus Unum*, 'one out of many'. In a speech given in Saint Louis in 1919, Theodore Roosevelt claimed that 'We have room for but one flag, the American flag . . . We have room for but one language here, and that is the English language.' Language was perhaps the most insidious tool in swallowing up cultures that still determinedly clung to their home ground.

Over two centuries after the first colonizing efforts in America, the suppression of the country's indigenous cultures was still sufficiently notorious for Charles Lyell to note that 'few future events are more certain than the speedy extermination of the Indians of North America and the savages of New Holland in the course of a few centuries, when these tribes will be remembered only in poetry and tradition'. As brutal wars and disease ravaged the Native American populations, the English language usurped the linguistic culture of the country. A host of Native American languages, like Cayuse, from Oregon and Washington, and Adai, from Louisiana and Texas, became extinct after the arrival of Europeans on the continent. Of the several hundred indigenous languages that remained, most disappeared in the following centuries or became increasingly endangered. In late 2008, the Native American language of the Eyak tribe in Alaska died out as Chief Marie Smith Jones, the last speaker, passed away at the age of eighty-nine. One of the main reasons for the demise of her language was an earlier policy to remove children from their native enclaves, in the misguided belief that indigenous children benefited from immersion in the dominant culture – efforts to make the children 'American' in the anglicized sense. But the final nail in the coffin was the dominance of the English language.

In 1553, Thomas Wilson, who went on to become privy councillor in the government of Queen Elizabeth I, published *The Arte of Rhetorique*, which lampooned the use of medievalisms or archaic coinages from Greek and Latin. Wilson argued for the virtues of the 'King's English', a plainer, standardized language. At the close of the century, the poet Edmund Spenser composed his *View of the State of Ireland*, expressing prejudices common during his era that the Irish were a rude bunch in need of domesticating through the English language. 'The Irish,' he wrote, 'were a people altogether stubborn and untamed.' After offering over two million acres of Irish land ripe for cultivation to willing investors, the English government facilitated so-called 'plantations' of English settlers to further their rule. Spenser became secretary to the governor of Ireland and oversaw the suppression of the Irish language. He was a devout believer in the Roman concept of imperialism: 'It hath ever been the use of the Conquerour, to despise the language of the conquered, and to force him by all means to use his.'

British imperialism brushed Standard English across all the countries that came under its rule, which helped it to acquire people, workers, land and resources. The British East India Company began trading in Surat in 1612, and later in Madras, Bombay and Calcutta; in 1835 Lord William Bentinck proposed an English education system in India. Similar systems would follow in Australia and New Zealand, in parts of Africa, and in Singapore.

Industrialization significantly expanded the language, when trade with Britain required those ambitious to share in new technologies to have competent English. With the demise of the British Empire in the 1950s, deliberate plans were laid to maintain political influence through the use of English. As the annual report of the British Council made plain in 1961, 'Teaching the world English may appear not unlike an extension of the task which America faced in establishing English as a common language among its own immigrant population.' Huge sums of government money were spent to cultivate English around the world, efforts matched by America's even greater influence. In post-war Europe, American foreign policy administered so-called 'America Houses', which propagated aspects of American culture and held language classes in English, along with the distribution Europe-wide of a considerable number of books printed in English.

By the close of the twentieth century, Standard English had become the first language in history to be spoken globally. In some ways, this ideal of a single tongue was laudable and necessary in an age of exploration, but English was a language haunted by those it had subdued and destroyed in gaining this prevalence.

These days, the mass media, most especially those of the USA, disseminate the language throughout the world. English was the first language transmitted by radio, in 1906, and now the dominance of American media products, from film to television to pop music, bombards other countries with the language. English was the cipher of the earliest forms of the Internet, through American Standard Code for Information Interchange (ASCII), and now Standard American English has superseded Standard English, transported across the planet to television sets in all corners of the world, downloaded or viewed online.

While English acted in some countries as an empowering and fairly unbiased language, a neutral tongue for communication across borders and ideologies, more often it was steered towards business. Jean-Paul Nerriere, former vice-president of IBM, which made the first commercial computers, argued passionately for 'Globish', a 'radically simplified version of living English'. Nerriere felt that business meetings internationally were 'dependent on idioms, trapped by culture-specific knowledge, hampered by a huge vocabulary'. He wanted a stripped-down version of English for simplicity, something fated to become the global dialect of the new millennium. He recognized that the world was overcrowded and that more people than ever would migrate to cities in search of greater prosperity. In these populous human landscapes, there was little need for diverse languages.

Darwin suggested that language focused the human mind, gradually augmenting the brain. His contemporary, the German linguist August Schleicher, argued in *The Languages of Europe in Systematic Perspective* that languages were akin to natural organisms. According to Schleicher's system of evolution, humans once existed in languageless depravity. The more people considered the origins of linguistic ability, the more anxious they became about a time when man might not have possessed the means of communication. In 1866, Darwin's cousin Hensleigh Wedgwood wrote *The Origin of Language*, which debated whether man in his superior form devised language for his own benefit or whether its origins lay in the shadows of a time when our ancestors were barely human at all. Wedgwood suggested that the imitative nature of language would arise from the early need of primitive man to emulate sounds in the natural world. Next, language would progress through stages of increasing differentiation, until it was endowed with the complex significance of modernity. In Wedgwood's theory, the nearer a word was to the natural world, the more likely it was to change. Only in the society and minds of modern man could the meanings around which morality and belief revolved be firmly established. 'In savage life,' he wrote, 'when the communities are small and ideas are few, language is liable to rapid change . . . When Indians are conversing among themselves they seem to have pleasure in inventing new modes of pronunciation and in distorting words.'

The idea became entrenched that the languages of indigenous people living closer to the natural world were cruder than those of the industrialized nations. Such pitiable speakers should be encouraged or forced to abandon their native languages and to revel in the greater opportunities afforded by 'civilized' speech. In truth, the most grammatically complex languages of the world were usually the more isolated tongues of those who were wholly dependent on their landscapes. A language spoken by generations of a small population spliced to a single site or activity could accrue great depths of information about their environment and foster a richness of expression. Proto-Indo-European, the original from which

English derived, was spoken across the Pontic–Caspian steppe of north-east Europe; it was far more complex than the languages that evolved from it, with twice as many tenses, voices and moods as English. The various tongues that sprang from it diverged from one another as the speakers drifted apart into new territories, and then each language diversified further into regional dialects. Only with the extension of central governance did these regional dialects begin to fade, replaced by standardized languages that tended to simplify and lose a wealth of singular distinctions. Thomas Wright's *Obsolete and Provincial English* recorded the Yorkshire word 'black-blegs' for blackberries, the Lincolnshire word 'lareabell' for the sunflower, the Leicestershire word 'mudgins' for the fat around the intestines of a pig, the Gloucestershire term 'wanti-tump' for a molehill, the Norfolk words 'swallocky' for the appearance of clouds before thunder and 'famble-crop' for the stomach of a ruminating animal, and the Cornish word 'drong' for a narrow path. Wright saw it as his vocation to translate the foreignness of the past.

Many language conservationists claimed that a world died with the extinction of a tongue, an idea derived from the work of the American linguist Edward Sapir, who argued that 'The worlds in which societies live are distinct worlds, not merely the same world with different labels attached.' Other linguists claimed a language did not capture specific knowledge and meaning that others could not. But did it follow that the transfer of information was as efficient if people all spoke the same language?

William Chauncey Fowler, a nineteenth-century author from Connecticut who wrote extensively on the English language, noted, 'As our countrymen are spreading westward across this continent, and are brought into contact with other races, and adopt new modes of thought, there is some danger that, in the use of their liberty, they may break loose from the laws of the English language, and become marked not only by one, but by a thousand shibboleths.' Similar, uncontrollable diversity pervaded the spread of English throughout the British Empire. While the British regarded the jungle of languages and dialects in Africa as unfavourable, a chaos over which only the civilizing tones of English might rule, the English that native Africans eventually spoke betrayed the patterns of their earlier speech and co-opted local words, shaping the tongue to express their private experiences.

The ghosts of earlier languages breathed in the idiosyncrasies of colonial English, as speakers often refused to relinquish their native language completely. The structures of their first language underpinned and altered their adopted tongue, preventing wholesale homogenization.

New York, like all other large cities, generated its own diversity through the energetic fusion of so many different people, expressed everywhere in great flowerings of colourful and insubordinate graffiti. This diversity was already evident in the nineteenth century, when the exuberant collision of immigrants and Americans in Manhattan bred a host of different dialects. Over time these were concentrated into

so-called New Yorkese, 'a thing composite and strange', as the critic William Dean Howells noted wryly in 1896.

I caught the train from Penn Station early one morning, passing out of the city and through the snowy Adirondacks that stretched all the way to the border with Canada. The air was dry as whisky on my tongue as I left the train in Montreal. I stayed in a hotel next to the bus station, a featureless block the colour of cold porridge. Lying in the room, I listened to the unabating growl of the bus engines. The following morning in the deep sunless blue of the winter light, I took a plane operated by an Inuit company called simply Air Inuit. Thirty or so passengers embarked, mostly Inuit women wearing white or pale-blue amautiit, hooded winter capes. A few Canadian stalwarts in fur-lined parkas sat at the back, softly snoring.

The cold greeted me first as I alighted from the plane at Iqaluit airport. In the waiting area, young Inuit women wandered about, their babies bobbing inside their hoods, ready to greet relatives. A small gift shop sold traditional Inuit carvings in soapstone or caribou antler of archetypical Arctic images such as polar bears, narwhals, or dogsleds, alongside cans of fizzy drinks, chocolate bars and international glossy magazines. Outside, the winter light had a moonlit otherworldliness, people trudging through the snowy streets casting cobalt shadows. The taxis, like an expensive bus service, were crammed with customers and charged a set rate per person for journeys. I flattened my puffy down jacket and pressed myself into a car beside three others. After each passenger had briefly stated their destination, we sat in silence and stared out of the windows.

'The city's alive with anything's-possible,' Mayor Sheutiapik was quoted as saying on the website of Canada's newest and fastest growing city. 'Even though it's located on the remote Arctic tundra, Iqaluit aims to be every inch a capital city, with the amenities and quality of life to rival any in Canada.' I had read this the night before flying here. It was ten years since the inauguration of Nunavut, which encompassed the far northern regions of Canada, and telephone cables strung the capital city like improvised washing lines, while four-wheel-drives and trucks fogged the streets with grit. We passed a supermarket called Northmart, a huge, characterless, rectangular building. People hung about outside its entrance, smoking or waiting for taxis. The supermarket was briefly notorious when a photograph appeared in the world press of two Inuit children sleeping rough between the bins and the concrete walls.

I was staying in Apex, a small settlement just outside the city, once occupied solely by Inuit when Iqaluit was an American airbase. In those days, Iqaluit was known as Frobisher Bay, and only regained its original Inuktitut name, meaning 'place of many fish', after the Inuit were granted control of the Nunavut territory. Until then, American policy forced the local Inuit to live in Apex. As we reached the brow of the hill, I could see the multicoloured lights of the hamlet against the bluish snow. I felt a little melancholy. I had been in two minds about whether to come on this journey. I was trying to limit or avoid flying, and certainly not to take such potentially destructive technology for granted. In the light of my qualms, I had written to one of my colleagues in Iqaluit to let him know that I would not be visiting the region after all. He took me to task on the issue. Bert was a retired Canadian teacher who had been living in the remote north for decades. He was a passenger on the first Boeing 737 flight from Iqaluit to Montreal in February of 1969. 'Your decision,' he wrote in an email, 'leads directly to you being unable to travel to our location to see first hand the changes you want to research in the Canadian Arctic . . .' The implication was, 'Well, I live where food and medicine and dozens of other items are easily available, but folks who live in remote and isolated locations should not be using jet airplanes and hence are denied access to these items,' and hence, 'They should not be living in such remote locations.' I responded in good humour, agreeing with Bert that one's ideals often rang hollow.

Still, I agonized over the decision for weeks, finally resolving to travel to Nunavut but to do as much of the journey as possible across land or water. Shortly before departing, I read David Mackay's brilliant *Sustainable Energy Without the Hot Air.* According to Mackay, a Cambridge physicist and advisor to the British government, to travel in a less damaging way, I ought to travel slower and travel less. A bike or my own feet would be best. Otherwise, a steady, slow train. I felt vindicated on those fronts, but he argued that the 'sad truth is ocean liners use more energy per passenger-km than jumbo jets'. Elsewhere I read that pumping out energy directly into the upper atmosphere is considerably more destructive, which complicates a simple emissions comparison. I travelled by ship on this basis.

I arrived in the dark reaches of the Arctic in December 2009 as world leaders gathered in Copenhagen for the United Nations conference on climate change. From a great distance, I read about these events as they unfolded. The Copenhagen Accord emerged from the proceedings, an agreement recognized by 193 nations to limit the rise in global temperatures to no more than a few degrees above pre-industrial levels. One of the special advisors to the Canadian environment minister was the National Inuit leader, Mary Simon. She felt frustrated that the Accord assumed climate change had an equal impact across the Earth's regions, regardless of the relative vulnerability of different habitats. She was disappointed, she said in her official response, that the lengthy debates in Copenhagen failed to recognize the disproportionate effects of climate change on areas dependent on ice and snow.

Shortly after I had settled myself at Rannva's bed and breakfast on Angel Street, a couple of other guests arrived. Sitting around the fire together, we talked about why each of us was here in the far north of Canada. One of the men, a geologist, was in the country to appraise the territory's mineral resources for a private company. I asked him politely if the profits from the resources would go to the Inuit people and he shrugged. 'To be honest,' he said, 'at the moment it's kind of a free-for-all. If you buy a licence, you'll get profits.' The following morning, I walked into Iqaluit in the tracks of countless skidoos, occasionally pausing to observe the sea freezing across Frobisher Bay, mists dancing upwards like chiffon in a breeze. I fell into step behind a man less muffled in polar clothing than me. Observing my furry hat, he smiled and asked, 'What brings you here?' I told him that I was in the city to research the changing relationship of the Inuit to their landscape, and he studied me, a little warily. 'Nunavut's future won't be the land,' he said, unprompted. 'It'll be funded by minerals.' Four years ago, he explained, the whole of the Meta Incognita peninsula was surveyed, confirming reserves of iron, lead, gold and diamonds. 'The more the ice melts,' he called over his shoulder as he turned away, 'the more they'll get.'

I was on my way to meet Meeka, an Inuit hunter with a Scottish ancestor, a whaler from Peterhead. I got lost and had to ask a woman if she knew Meeka. She pointed to a large brown house with an Inuit sled-dog chained outside and a caribou skin hung from the wall. Inside, the house mingled

Inuit objects and modern appliances. It was warm, lived-in and wonderfully inviting. Caribou flesh was drying on a rack near the table and some strips of the russet meat lay before Meeka. She'd made me a pot of English tea. We exchanged pleasantries, each in gentle greeting of a stranger. I told her about my thoughts on seeing Northmart and she made a moue of ambivalence. 'I am the first generation to have seen things like pineapples and oranges,' she said, her eyes full of humour. The Inuit shared food according to very strict methods of harvesting and storage, she explained, and they would never give unhealthy food to one another. People were trying to adopt the same attitude to the supermarket food. But Meeka resented outsiders placing moratoriums on the Inuit hunting wild animals. 'Where is the moratorium on the circulation of products like olive oil?' In Meeka's view, the profligate way of life that allowed the worldwide distribution of products did more to endanger wildlife than the controlled hunting of the Inuit, although she recognized that this was contingent on their small population.

I began by telling Meeka why I had come all this way to see her. I confessed my own deep sorrow at knowing so little of the natural world, and at my lack of a sense of belonging to a particular place. After a long pause, Meeka began to speak. For fifteen years, she told me, she had been teaching 'land-skills' in schools, attempting to repair her people's previously unbroken inheritance of knowledge of the Arctic landscape. Despite her best efforts, though, she confessed that over the last five years, the basic knowledge of the children

had considerably disintegrated. When she took teenagers out on field trips, she realized that they didn't even know how to find water – 'the most basic resource' – a sign that their generation was very far behind now. As a hunter, she explained, one must first learn how to detect a path through a landscape, leaving it almost unmarked. Inuit hunters used hints in the ice to piece together their route. And an individual had to know how these signs changed through the seasons.

A slight woman with round, insightful brown eyes, Meeka still hunted throughout the year. Her knowledge of the landscape began at birth and was inscribed in her earliest memories. 'By five, a boy should take his first seal,' she told me. In the Inuit tradition, children would carry the meat home from a hunt so that they learned not to eat it for themselves but to share it among the group. By living closely with the landscape from the first years of life, the children gained an almost instinctive sense of boundaries. This gave rise to strict laws for stewardship of the land, although she clarified that they did not perceive them as laws. 'You do not harvest from the nurseries of the various animals,' she said. 'You have signs of the wellbeing of each species, the size of the caribou. You never shoot the "scouters" – the first caribou that appear. You wait for the herd to arrive. You never speak negatively of the animals or you upset the animal spirits. You assume that you can take what you need and no more, and as much as you need, and as long as the spirits are happy, the supplies will be bountiful.'

When the Canadian government began relocating the Inuit

who lived out on the land, and imposed different methods of schooling, the traditional means of passing on knowledge inside Inuit families collapsed. Until the 1970s, the schools to which Inuit children were sent didn't teach their native language of Inuktitut. The aim of this, Meeka felt, was to change their perception. 'They wanted us to see ourselves as individuals,' she said, 'but we didn't think of ourselves as individuals. That's not the Inuit way. We teach our children to care for one another as part of a community.

'My daughter was traumatized as a child,' she told me. 'I was alone and had to hunt alone, and, as a single mother, I had to take my child with me in the coldest months, when it's minus thirty, minus fifty, minus sixty, and it's frightening for a child . . . She is not confident hunting now. I should not be a hunter alone. We hunt together, we need others. As a community.'

As our conversation began to flow more naturally, Meeka spoke of her first experience of a big city in the early 1980s. She was flabbergasted at the sight of a homeless person. She couldn't understand why nobody was helping her, and why she had no family to take her in. 'I stood with her for so long, until she herself stood and told me, "It's OK, I'm fine," and then I felt I could leave her.' It was because of these sentiments that the photograph of the two runaway children outside Northmart caused such shock and discomfort. Meeka feared that children now had no hope. They harboured unrealistic expectations about what the world would provide for them.

She gestured towards the rack of drying caribou. 'Would you like to try some?' she asked. Not being much of a meat-eater, I agreed only out of courtesy, chewing with trepidation. In her spare time, Meeka recorded the elders, the last generation of Inuit unaffected by government interventions. While the education of the younger generations took place further from the home, she hoped that by documenting the profound knowledge of the elders, she might halt the erosion of their unique world. Crucially, the words of Inuktitut, their language, cupped their gift for sustaining themselves from the land. The English that we spoke together that morning was inadequate to the task. English was a stranger to the Arctic landscape, and the perceptions of its vocabulary would blind someone's way on the ice.

'Let me tell you about Inuktitut words for "ice",' Meeka said, 'all the different stages of ice. They needed to be specific to the regions of Inuit, as ice behaves differently in different places.' There is *aniksaq*, large pieces of floating solid ice with no cracks, separated from the floe edge by strong currents or winds; *apputainaq*, the false ice of new cracks covered with snow; *iniruvik*, ice joints occurring along the thinner edge of what was the floe edge, continually opening and closing; *kiviniq*, the depression usually formed near shorelines by the weight of tidewater rising through cracks; *nappakuit*, ice only a few millimetres thick when broken by the forces of winds, currents or waves; *siatuninik*, pieces of ice moving as a group in the current; *sikuliak*, or newly formed ice with no snow on top, thinner than old ice, but safe to walk or travel on.

She then fetched her computer and showed me a PowerPoint presentation designed for use in schools. The pages illustrated the Inuktitut words for the nine stages of a polar bear's life: a yearling, a three-year mother (which means the mother has two other young), and so on, none of which had an English equivalent. Next, she showed me the complex and variable methods by which the Inuit interpreted the seasons of the year: some were essential and yearly, and others surmounted the fluctuations of the years, allowing for aspects of a season to arrive at different times. She spoke of the knowledge packed into the economy of their words, how the word for 'wolf' in Inuktitut is *aqalik*, or 'one-who-carries-it-on-its-back', because the wolf bites its prey on the right side and throws it up on to its back between its shoulders. 'The English word "wolf" says nothing. It's just a name,' Meeka said. As more and more Inuit children spoke English as their preferred language or became bilingual, Inuktitut was losing some of its nuances. As the lives of this younger generation drew further from the landscape, the memories of the land captured in the utterances of Inuktitut dimmed.

Our conversation was drawing to a close. I did not want to take much more of Meeka's time. Despite her steady, soft-spoken manner, I knew that she wanted to go to her dogs and feed them. But before I left, she told me about a sentiment she called 'yearning'. She drew this feeling for me. 'This is our yearning to return to a close association with the land.' She gazed at me unfalteringly. 'In the South,' she said, 'you see things . . .' and she drew a straight line along the table. 'Inuit see things . . .' and she drew a coil on the table. She pointed to a place at the beginning of the coil, and then to a place further along it, where her slender brown finger paused. The coil, she said, stood for all the generations' knowledge, and this yearning was a desire in the present to return to that place at the start of the coil, which stood for an experience or perhaps a place in the past. Yearning was the means for the past and present and future to amalgamate in a single sensation.

The temperature was about minus twenty and the air scratched my throat, alarming my blood as it entered the dark warmth of my body. Everywhere was the sound of traffic, the pips of vehicles reversing, the bark of tempered acceleration. My initial impression of Iqaluit was of an awkward blend of industrialization and Inuit culture. Some aspects of modern industrialized life were assimilated into the environment simply and successfully, while others seemed utterly incongruous against the tundra.

Strings of mauve cloud like footprints in the sky headed into the unstirred landscape that lay just beyond the borders of the city. I walked out to the shoreline, past the gift shops and an enormous hotel. It was three o'clock and what little light suffused the day was already softening into twilight. Across the anarchic structures of the frozen sea, I made out a line of snow-covered hills banked by clouds of a deep turquoise where the tiniest sliver of orange hinted at the expansive sunsets of elsewhere.

The twelfth-century *Islendingabok* recorded the first discovery by European explorers of human habitation in the
Arctic. They found fragments of boats and stone artefacts
belonging to the so-called Arctic Small Tool Tradition (otherwise known as Pre-Dorset), a circumpolar culture dating
from around 2500 to 800 BCE. A more temperate climate
possibly facilitated migration and the spread of this hunting culture, characterized by small, delicate weaponry. Some
of these hunters somehow evolved their skills sufficiently to
escape the shadows of the woods and to begin life on the ice.
Splinters of bone and small stone projectiles, telltale rings of
campfires and the indents of shelter – such remains left a fragmentary picture of the life of small groups of hunters. These
people exploited each season's bounty of seals, caribou, fish
and geese, and made their homes of bones or driftwood
draped in animal skins. After 1500 BCE the climate cooled
again, and the behaviour and range of plants and animals
shifted in response. Some indigenous tribes adjusted their way
of life accordingly. There is a theory that Pre-Dorset hunters
migrated into the deserted regions, returning to old camp-
sites over several seasons, pitching their tents around a central
fire. In the starless chill of the winter months, they hibernated
in their beds, stacking up snow for insulation. Surveying the
huge frozen wilderness before me, I found it almost impossible to imagine how anyone could survive out there.

After 800 BCE, there is no further archaeological evidence of these people and their distinctive culture, but they
were succeeded by the Dorset tradition, people with more

sophisticated armoury engineered to hunt from the sea. They had polished chisels, sharp blades carved from slate, and advanced harpoons. They took geese and seals, along with beluga whales, narwhal and polar bears. They had beautiful ivory snow knives, and sled-runners were found at sites on Baffin Island, evidence that they pulled sleds by hand. Over a thousand years, this culture was superseded by a hunting tradition advanced enough to harvest bowhead whales. The Thule people hunted for large sea mammals in open water using skin boats, capturing their prey with a harpoon line and drag floats. They used dogs to pull their sleds. They lived in winter snow-houses, entered by a subterranean tunnel – cramped homes lined by rocks with narrow sleeping platforms. In the frozen months, they lived off the surplus from the summer bowhead-whale hunt. From spring onwards, they inhabited tents of animal skins, the rafters made of the ribs and jawbones of whales, the distinctive homes with which the Inuit became synonymous. They spoke varieties of Inuktitut and lit their evenings by burning blubber oil in soapstone lamps. The harvest of a few bowhead whales by each group allowed a higher proportion of children to survive. Cramming their belongings and children into their characteristic boats, a family group would go in search of a whale, engage in the hunt, and swiftly raise a new camp amid the slaughter.

When the Arctic began to cool again, winters became interminable and starvation nearly plunged this culture into extinction as well. Those who survived did so by making the most of any sustenance, hunting an array of wildlife throughout the year. Through these hardships, diversity arose across the Arctic, each group adapting their skills to the specific opportunities of their region. The Inuit armoury became highly specialized, with toggles to drag the dead weight of a kill across the ice, float harpoons, snow goggles to protect their eyes from the glare, ivory pins to plug wounds to keep the meat of a dead animal fresh. Peering across Frobisher Bay and shivering, even in bundles of synthetic polar clothing, I marvelled at a society that perfected subsistence in this bitter, glassy territory.

Immanuel Kant's *On the Dynamically Sublime in Nature* considered the moral implications of people living in remote and frozen regions. He cited such forbidding environments to counter the common view that the natural purpose of the Earth was to serve humans who were ordained to exist in the landscape. As it was by no means evident why people should live in these inhospitable places, 'it would be hazardous and arbitrary indeed,' he concluded, 'if we judged that vapours fall from the air as snow, that currents in the sea bring timber grown in warmer lands, and that large marine animals replete with oil are there because the cause providing all these natural products acts on the idea of an advantage for certain wretched creatures'. I, too, felt that these frozen tracts served to remind any observer of the origin of humanity inside nature, the self-importance of our species dwarfed by the magnitude of the ice.

But Baffin Island, like the rest of the Arctic, was recasting itself. Temperatures seemed to have been increasing since the beginning of the twentieth century, thawing faster than nearly anywhere in the world. Some species of Arctic midge, which thrived in the glacial conditions, had already become extinct. A range of other native wildlife would undoubtedly follow. Now that human activity, not the Earth's orbit around the sun, was the chief determinant of the temperature in the frozen North, the chastening promise of the ice was deteriorating.

I headed away from the shore to the government building to visit Leesee, another hunter. The woman who greeted me seemed at once shy and self-assured. During the day, Leesee worked for Parcs Canada but she still considered herself a hunter. Her father grew up out on the Arctic tundra near the village of Pangnirtung. He witnessed starvation and hardships that were an inevitable feature of survival in such forbidding natural conditions. These experiences convinced him of the importance of remaining versatile. In contrast, Leesee said, her mother 'had her nose in the air'. She had grown up in relative prosperity in Pangnirtung and had never known hunger. 'This made her more obstinate,' Leesee confessed, 'and less comfortable with change.' Her father encouraged his children to go to school but he also educated each of them in the traditional ways of Inuit culture and counselled them through his own knowledge and experience. The contrast between Leesee's parents suggested an intriguing reversal whereby proximity to nature, rather than impeding progress, made people more convinced of the need for change, more aware of their fallibility. Those in their safe houses insulated

from the severity and risks of the landscape were more in-
clined to provincialism and a lack of foresight.

Although she lived in Iqaluit, Leesee looked out of her
window each morning at the weather, as if she was still out
on the land somewhere. 'I notice even a change in the dew.
When there's a dew on the snow, I know it's a different kind
of snow.' She knew that this attentiveness was the ghost of the
earlier obligations of her forebears. Her generation no longer
needed to survive in the same way. While she worked for the
national parks in the hope of spending as much time out on
the land as possible, she also consulted weather forecasts to
check how conditions might change, and availed herself of
guns and a snowmobile. Her own children, brought up ex-
clusively in the city of Iqaluit, now showed little interest in
traversing the tundra, finding it too cold. Leesee didn't wish
to dictate to them who they should be or what they should
strive for in life. But she still hoped they might find their way
back to their native landscape. I asked Leesee if she, like her
father, had passed on her knowledge of the land. 'Do they
know as much as me? No, they don't,' she admitted quietly.
'But they're still raised with Inuit ways and I hope they will
find a middle road.'

The airstrip cleaved apart the village of Pangnirtung. Although a small settlement, it was brash with the sounds of
constant traffic and children streaking along on skidoos. I
was staying with Markus, a German nurse who had worked
in the community for decades and kindly offered to put me

up. A large patio window in his house captured the extraordinary beauty of Pangnirtung fjord. On my first evening, I walked through the town in the navy light of midwinter. Two young boys were amusing themselves by sledding down a small hill. Seeing a stranger, they came running up to me, smiling and waving. 'Hi!' they said. They were delighted to find I was English and immediately offered me a ride on their small plastic sled. I declined, laughing, but the boy that spoke the better English called after me, 'Will you watch us?' And so I stood at the bottom of the hill, applauding as they raced down higher and higher slopes.

The town had a couple of general stores, a museum and a craft centre, as well as the visitors' centre for Auyuittuq National Park, a large expanse of ice fields and tundra, home to Arctic hares and foxes, ermine and polar bears. Skirting the shoreline, I came across a white dilapidated building with a sign that read HUDSON'S BAY COMPANY OLD BLUBBER STATION. Two slender whaling boats were decaying into ribs of wood out the front, each painted white with a rim of blue and an underbelly of red. Someone had dumped a child's bike in one of the boats, while the other lay upturned, rotting in the snow like the carcass of a whale. Around the back of the building were thirty or forty rusting drums for whale oil and a large vat and pump.

The Hudson Bay Company established its base here in the 1920s, but the Scottish whaler William Penny reached Cumberland Sound, which stretched just beyond Pangnirtung fjord, as early as 1840. Penny had been a whaling captain

for only five years and he had witnessed dozens of British ships being wrecked in the Davis Strait. Coming up against extreme ice conditions, whalers found themselves locked in the Arctic with diminishing supplies. They lacked the skills to hunt for food and their investors forbade them to eat the whale catch. Occasionally, they were in such dire straits that rescue parties set out with supplies and manpower, including a fleet led by the British explorer James Clark Ross, who sailed for the Arctic in 1836 in his ship the *Cove*. Penny was determined to improve conditions for the whalers and became convinced of the need to remain in the Arctic through the winter months, establishing stations on the shore. He had heard rumours of a region with plentiful supplies of whales that promised to be a favourable site for his plan. In 1839, an Inuk called Inuluapik drew a map of the Cumberland Sound, and this new hunting ground opened up for the European and American whalers.

Inuluapik sailed to Scotland with Penny the same year,

inhaling both the scent of trees, which he found wonderful, and the tuberculosis that eventually killed him six years later. A large whaling station called Kekerten was established around fifty kilometres from Pangnirtung, now a ruin of rusting metal, old homes, spoiling vessels, and graves. It was one of the central sites of British whaling in the Canadian Arctic and Penny's boat was still crumbling to dust on its shores.

'In 1846,' wrote one whaler, 'which was the first year I went to Cumberland gulf . . . we was terrified to go out in the boats, the whales was that large and numerous they raised quite a heavy sea with their tails and fins.' By the 1860s, foreign whalers had plundered the native population of bowheads. They supplemented their catch with seal skins, fox furs and walrus tusks. They also struck bargains with the Inuit people drawn to their settlement, employing them during the freezing months. The American anthropologist Franz Boas studied the relationship between Inuit and whalers in the mid 1880s. 'When the Eskimos who have spent the summer inland return at the beginning of October, they eagerly offer their services at the stations, for they receive in payment for half a year's work, a gun, a harmonium or something of that nature, and a ration of provisions for their families, with tobacco every week.' The Inuit hunters were eager for weapons and tools that would help them in their struggle to live from the land – knives, drills, rifles, saws, metal pots and so on – and discovered the addictive pleasures of sugar and tobacco. Mathias Warmow, a missionary who travelled to Kekerten with Penny in 1857, viewed the Inuit caught between

their old ways and European influence as distorted in their natures. 'I am always sorry to see the Esquimaux imitating the Europeans in all respects . . .' he wrote. 'They were undoubtedly better off in their original state, and more likely to be gained for the kingdom of God. But when they begin to copy our mode of life they are neither properly Europeans or Esquimaux, and will speedily die out in consequence of the change.' For Warmow, a nature of neither one place nor the other appeared to him as a deviant form, a disposition he couldn't associate with as a peer or inveigle as a perceived innocent.

Climbing the hill out of the town, I giggled as I repeatedly sank up to my thighs in the deep snow. After a reasonable distance, I slumped down in the coldness and looked out across

the view. The blue of the night world shaded from the dark azure of the snow to the motionless midnight-blue tongue of the fjord to the indigo of the exposed rocks of the mountains. Most mesmerizing of all was the full moon; suspended in a gap between two sheer-edged mountains, it was palest yellow, perfect and harmless. The snow glittered on the ground as I twisted my head to catch sight of it, as if the moon had shattered into a trillion pieces. I could see all of Pangnirtung, which now seemed insignificant amid the enormity of the untouched landscape. The refuse dump smoked at the far end of town and I could just make out the skittering sled-dogs on their chains at the outskirts. Someone in Iqaluit had told me that few Inuit still used the dogsled to hunt. They kept dogs, now, for the tourists who came in the spring and paid thousands to travel with them in the wilderness.

In the Inuit imagination, the idea of nature remained incorruptible and, in its own way, inseparable from human existence. The Inuk's kinship to this natural world derived essentially from use of it, and this successful exploitation was integral to an individual's sense of character and worth. For the Inuit, their affinity with nature was still traceable to their indispensible need for it, an aboriginal longing for a landscape of such dramatic importance to them.

Reflecting on Leesee's recollection of the differences between her parents' temperaments, my thoughts returned to the suicides that afflicted Inuit communities. I wondered if the loss of this affinity with the landscape made the shockingly high rate of suicides among the Inuit youth, particularly

young men, more comprehensible. Over the last decade, the government and community groups had tried to tackle the issue, but a definitive explanation for the phenomenon had not emerged. In the last year, there had been eight suicides in Pangnirtung, two already in the previous month, a remarkably high number for a settlement of only a thousand or so people. As part of the process of gaining self-reliance, Inuit playfully taunted their children, making jokes about abandonment – games provoked by the formidable dangers of the Arctic. Settled, modern life distorted such parental practices. Many Inuit children who had attempted suicide claimed they were bored. One of the prevailing theories holds that the young exist in a cultural hinterland, denied a comprehensive experience of their native tradition, yet separated from the dominant culture to a degree that prevents them from fully adopting its skills and values. Like everyone else, I wondered what filled these young people with such despair that they terminated their lives. Perhaps there was an irreconcilable conflict between the yearning that Meeka had described and the new cultural landscape in which they were raised. After descending the hill, I passed the frozen water-supply reservoir on which children had clearly been skating. I caught on the wind the voices of children playing together, teasing and lively. They were no more than eight or nine years old, and they seemed incorruptibly happy.

The following morning, I went to the Park Visitors Centre to arrange meetings with some members of the community

who grew up out on the tundra around Pangnirtung. Billy Etooangat, a playful, blue-eyed Inuk who worked as the interpretation officer at the centre, greeted me at its entrance. Like Meeka, he felt that the Inuit culture was changed fundamentally by teaching the young at school. 'I was sent to a school where I learned my A-B-C and I was taught to get my qualifications, so I could get a good job. I got a good job, right?' he said, shrugging his shoulders and challenging me with a smile. As we talked about my work, he explained that everyone lived in traditional camps on the land until the 1930s, when gradually people shifted to the settlement, and that, in the more distant past, these camps were scattered throughout the region, each containing large whale bones. 'They knew,' he said, 'how to get close to these large animals so that they could hunt without modern weapons. But that skill was lost to the past.' Billy was only half teasing me when he blamed the undermining of traditional Inuit knowledge on the creation of urban settlement. 'Pangnirtung is artificial,' he said. 'In the town nobody can listen to the elders, there's so much noise.'

In the early twentieth century, Billy's grandfather worked for several seasons as an oarsman for the European whalers. By this time, Billy told me, the whales were scarce and the traditional Inuit ways of hunting the creatures were not being passed on to the new generations. 'I learned to hunt with a gun,' he said. Billy suggested we watch one of the centre's films about the bowhead-whale hunt. In 1996 and again in 1998, the community made efforts to renew the

hunt that their forefathers undertook. The bowhead whale had become so endangered that the Canadian government had enforced a nationwide ban on whaling. Following a successful campaign, the Inuit hunters of Cumberland Sound were given a licence to hunt one whale. Although the catch was limited, the community saw it as a genuine endeavour to salvage the past and to close the loop between their lives and the nearly lost tradition of their ancestors. The first hunt was a failure. The four boats involved were launched at the same time and there was some crossfire. The whale that they captured sank, spoiling the meat. But the second hunt two years later was more successful. After several days, they managed to seek and snare a twelve-foot young male, using both old and new methods, and learning from the mistakes of the previous hunt. Each boat went out separately and took a shot. While the first two harpoons failed, the third harpoon struck and an Inuk called Simionie then fired the killing shot from a gun. 'Inuit believe their endangered whaling skills have been revived. Just like the bowhead,' one witness said. Speaking with pride at his son's achievement, Simionie's father confessed that only this modern gun would have killed the whale.

The nostalgic sentiment evidently pervaded Inuit life but should their re-enactment have inspired the revival of old methods? Or was Simionie's father right, and the technologies brought by the commercial whalers were simply better at the task? Billy had arranged for me to meet Simionie. He was a blue-eyed, pale-skinned Inuk, a large man with a kind,

friendly face. One of his ancestors on his mother's side was a Scottish whaler. Simionie was born in a camp peopled solely by his parents, siblings and cousins. They hunted every food source available to them to survive, seals and fish in the summer and caribou in the autumn, which involved migrating to their winter camps. Often, he said, they took nothing with them while hunting and relied on what they caught to clothe and house them for the next few days before they returned to the main camp. They knew the locations of all the old camps, both their own and those from the distant past. These remains, whether of their ancestors or of other groups, guided the Inuit across the tundra. The older camps contained the bones of smaller bowhead whales, which suggested that the traditional whale hunters chose to pursue beasts that were a more modest size and therefore easier to capture.

Simionie's knowledge of how to hunt came from accompanying his parents, early lessons in how to move, breathe and understand the landscape through shadowing his father as a child. For his part, he'd adopted a snowmobile for his hunts and felt nothing but the benefit of it. But some of the purpose had disappeared from Inuit lives: families were fracturing, unable or unwilling to share and needing to rely on one another less. On the outskirts of Iqaluit, surrounded by hangars and warehouses, was a prison. The majority of the inmates were Inuit men, incarcerated mostly for alcohol-fuelled acts of aggression and often for domestic violence. It was something that Meeka and I had touched on in our conversations. 'Violence against women is a real problem,' she

told me. 'It has always been a problem. Inuit are a jealous people. But now it is much worse.' Simionie blamed some of this on a kind of distemper experienced by people unwilling to adapt or with a sense that their adaptation was forced and unnecessary. He wished they'd realized twenty or thirty years ago how imperilled their culture had become.

'Life's easier and the older people are better cared for,' he said. 'Inuit have good positions in government now, and they got the language back in school.' A schoolteacher himself, Simionie also took the time to educate teenagers in the skills of survival out on the Arctic landscape, something he loved. He still preferred eating wild food and wanted to pass his passion on to the younger generation. But, like Meeka, he recognized that knowledge of nature had faded swiftly. More than anything, he feared the slow dying of elders like his parents, who possessed the last real knowledge of their ancient camp life. 'It's like a clock. You understand? Time moves on in one direction and before you know it, every-thing has changed. And you didn't even realize it.'

Watching the film of Simionie's hunt in the darkness, I wondered whether the Inuit of Pangnirtung had lost their belief in the traditional methods and skills, or if hunting in earlier times called for such a refined balance of dexterity and comprehension of the environment that it could be revived only with the complete knowledge of previous generations. Before I left, Billy called Juavie, one of the hunters who took people out on the land. I wanted to travel beyond Pangnir-tung and witness the conditions of the Arctic for myself, so

we arranged to go out to some of the sites of the old camps. I thanked Billy for his help.

'Your whalers all left in the early part of the last century,' he said. 'Left us our blue eyes too,' he added, grinning.

I sat in Markus's kitchen, bleary-eyed, tugging on a second pair of thick woollen socks. Although it was the early morning and half-light hadn't yet brightened the scene, an electric glow came from one of the sheds down by the shore. It was built of plywood, but shaped like the traditional whale-bone summer houses. One of Markus's neighbours was down there, building a new canoe or 'seeing another woman out of sight of his wife', Markus suggested ruefully. The little plywood shed looked so incongruous in the exquisitely fragile starkness of the Arctic landscape. The man might once have built the shed of whale bone; now, due solely to the commercial exploitation of whales by industrialists from other countries and cultures, a more-or-less universal ban on whale-hunting existed. Meanwhile, plywood was universally available, regardless of the destructive implications of its manufacture and transportation, even in treeless places like the high Arctic.

Perhaps, one day, the descendants of this man will witness the return of trees to these latitudes. The Arctic is melting; the remarkable variety of its wildlife will eventually disappear from the world. The Intergovernmental Panel on Climate Change, established by the World Meteorological Organization and the United Nations Environmental Programme, argued forcefully that the ice is now disappearing at such a

pace that the Arctic sea could be wholly ice-free throughout the summer months by 2020. Manufacturing companies
received the news eagerly that the Arctic trade route of the
north-west passage will finally open up, shortening some of
their shipping routes by as much as two weeks and increasing
their profits. The Canadian government will soon construct a
deep-sea port at Iqaluit, a large harbour at Pangnirtung, and
an inland waterway to support this new shipping route. The
countries bordering the Arctic – Russia, Norway, USA and
Canada – lap at the edges of the area that might contain a
third of the world's crude-oil supplies. If such industrialization in the Arctic materializes, for a time the bounty will be
enormous. The population of the Canadian Arctic had already
soared by 16 per cent since 2000. Tourists seeking authentic experiences of nature flew there in crowds each summer,
while a large number of cruise ships laddered through the
ice-free waters. During the brief rush to exploit its finite resources, the Arctic will swiftly transform from a quiet, largely
unspoiled place in which people have co-existed sustainably
with wildlife to one dominated by human presence.

Juavie came to the door and glanced at my outfit, a range
of expensive polar gear from an outdoors pursuits company
in Montreal. He said nothing, but Markus had already told
me, 'He won't let you get cold. But do what he tells you
and don't argue.' I understood the implications of his amiable
comments – the first whalers were woefully underdressed
in woollen caps and Guernsey jumpers, and their sense of
superiority led to several humbling deaths of starvation and

hypothermia. We were travelling on a traditional Inuit sled made of wood and layered in animal skins, pulled by a snow-mobile. As we left the town and headed across the fjord, I saw the dim yellow beams of other skidoos making early-morning hunting trips. Some disappeared into the purple darkness of the steep entrance to the Auyuittuq National Park, while others beat a trail ahead over the frozen waters and into the snow-covered valley beyond. The water of the fjord had solidified into an arthritic crookedness. The journey out to some of the old summer camps took us through a territory that was both monotonous and intricate. There were few signs of life. Scintillating whiteness dominated for thousands of miles.

After an hour, I felt cold overcoming my extremities. Juavie, as if sensing my discomfort, stopped the snowmobile and gestured for me to walk about to warm up. As I tramped over the snow, I was struck by a quiet the like of which I had never experienced. It was a still, snowless day, nothing moved, nothing sounded. It seemed like the world's grave, the breathless Earth downcast by life. And yet it was flawless and beguiling. 'Feet cold?' Juavie asked. I nodded. He rust-led about in the back of the sled and pulled out two hand-made fur boots. 'Caribou,' he told me. He took off my snow boots and stuffed the inner lining inside the thinner, flexible boots with soles made only of skin. I couldn't believe they would be warmer, but, as I walked about, heat gathered into the boot and I could wiggle my toes to warm them up. But his other gift of beaver-fur gloves left my fingertips cold, so,

when I felt it might go unnoticed, I slipped on my old mitts of synthetic fleece.

We didn't speak as we travelled. My mind moved erratically through thoughts and memories. I remembered Meeka describing how the youth of her people, who chattered in English and in a form of Inuktitut from which some of the brightness of this landscape had dissolved, now strung sentences between the old ways and the vocabulary of cities thousands of miles away. The wilderness across which we sped dwarfed the minor developments of Iqaluit and Pangnirtung. While its vastness still dominated, perhaps the Inuit would always feel the need to recall the knowledge and skills that allowed their people to survive in unbelievably fierce and unbroken habitats. The Inuit of Baffin Island who travelled across the sea ice understood the histories of autumnal ice, of land-fast and floe-edge ice, of familiar spring cracks, of the kinship between ice behaviour and tidal currents, winds, lunar cycles.

But all of these ideas and experiences might vanish if the places which engendered them alter. Once these frontiers are transformed, all recollection may also disappear of a closeness to nature that enabled some of our species to belong to one of the harshest habitats on Earth.

Shortly before Juavie had arrived, Markus had told me softly that Juavie's son had taken his own life a few weeks earlier. I'd looked at him in astonishment. Surely Juavie wouldn't want to take a stranger from England out for the day while consumed by the loss of his child? 'No, no,' Markus

replied, 'it will be good for him to do what he loves most, to be out on the land.'

Standing quietly beside Juavie as I did that afternoon, gazing out on the flat plateau of a frozen lake that had fed his ancestors for hundreds of years, I understood. When I told him that I wanted to see the old camps, he looked at me with a mixture of bemusement and confusion. 'Nothing to see,' he said. 'They're under the snow.' For Juavie, these camps only found purpose in his activity on the land. He wasn't yet detached from his native country in a way that might allow him to view it in a purposeless manner, like a sightseer. The ghost structures of the old camps beneath the frost came to life for him not as a spur to philosophical thought or regret but as markers of a life still fastened to the landscape.

Endings

Wicken Fen, Cambridgeshire

After returning from Antarctica, I went to live on a narrowboat on the River Cam in Cambridge. One morning in late
March, the river metallic in the spring sunshine, I was cycling
in a bit of a dream down Garret Hostel Lane. As I reached
the traffic lights and came to a halt, I noticed the sudden
persuasion of a purple crocus in the grass. It was a familiar
sign of spring, although not a native plant, and the first bloom
I'd seen after a long, cold winter. During the Middle Ages,
monks harvesting saffron for medicinal purposes are thought
to have brought crocuses to these soils from Mediterranean
Europe. While the traffic lights switched to green, I got off
my bike and took a picture of the flower. When I returned
home that evening, I phoned my husband, Ewan, and told
him I was going to move the boat out of town. I warmed the
engine, unhitched the mooring ropes and steered downstream
towards the Cambridgeshire Fens. I wanted to draw away
temporarily from the controlled environment of the city and
seek out a horizon constructed more haphazardly by nature.

I hoped to reach the moorings at Fidwell Fen by nightfall, and from there I would take the boat down the narrow
lode into Wicken Fen. Steadying the boat through Bottisham

Lock, I passed fields in which herds of cows lay humped, the bones of their large skeletons visible beneath their hide like windblown structures. Dusk refined the river and the countryside, hallucinations of the day's dying light. A heron, startled by the boat's wake, stowed its slender legs and took off from the riverbank, only to settle a little further ahead. As the boat travelled downstream, the grey silhouette of the heron rose into the air again and again like an angry ghost.

Beyond the neat mown grass of the mooring strip, there was a raised bank at the top of which a rough path the width of a man's shadow ran. Wild flowers, shrubs and grasses overtook the slopes, clashing in orderless harmony. On the other side of the bank were ploughed fields, the soil black and sumptuous as if farmers had stirred into it something rich, like molasses. A pitted road was visible between a row of trees and a ditch, leading to distant farm buildings. Through a small, rotting gate in the cool of several old crack willows, there were fields grazed by cattle, the ground boggy from proximity to the river. A group of birds were mirrored in a large pool of water on the edge of the first field. Wildlife unabashedly colonized the spaces left without housing or infrastructure, and yet human hands had contoured every part of this landscape.

As I walked through the fields in the dark, I noticed some pale flowers, their presence let slip by the moonlight glinting on their petals. I felt a sudden rush of delight. When I woke the following morning, I experienced a strong desire to discover the name of the flower. It turned out to be the

cuckoo flower. What motivated my pleasure and interest in this unknown flower in the darkness? This first fascination soon turned into a daily rite. Each day, I walked along the riverbank or in the fenlands solely to see what new flowers had bloomed or to pay attention to the fresh flurry of life. I made a note of each bird that appeared on the pool and of new generations of waterfowl hatching along the riverbank. I watched dragonflies mating in the reeds, balancing their sexual urges on the empty promise of midair. Gathering identification cards from the local bookshops and peering patiently through Ewan's binoculars, I endeavoured to put a name to all the life I saw, and especially to the wild flowers. I purchased a guide to British species and a notebook, identified anything I didn't know, and wrote down the month and site of each new discovery. By early summer, I was able to glance at a common flower in the hedgerow and, at the very least, recognize to which family the plant belonged. Nothing had ever made me happier.

In an essay on wild flowers, Richard Jefferies described the deep-rooted satisfaction he took in the flowering plants of his country as a child. It was a delight to him just to see them, he wrote. 'Without conscious thought of seasons and the advancing hours to light on the white wild violet, the meadow orchis, the blue veronica, the blue meadow cranesbill; feeling the warmth and delight of the increasing sun-rays, but not recognizing whence or why it was joy.' Jefferies' knowledge grew from a child's unselfconscious engagement with the world, just as my grandmother's did. His rural childhood

allowed his pursuit of knowledge to begin earlier; his first conscious thoughts were to discover the names of the flowers, he explained, but then he awakened to all that he hadn't previously noticed. The world suddenly flashed with detail.

Coming across his words in a secondhand book bound in faded blue cloth a year after my own first excursions, I knew exactly what Jefferies had felt over a century before me. It reminded me of Adam Smith's paper entitled *Of the Origin and Progress of Language*, in which he pictured 'two savages' coming together to dwell in the same place and dream up words for the features of their new, shared world. 'The cave they lodged in,' he wrote, 'the tree from whence they got their food, or the fountain from whence they drank, would all soon be distinguished by particular names.' Jefferies' appreciation of wild flowers was a thatch of experience, sentiment and distinctiveness, originating in the grounds of his childhood and upbringing. The memories of his home accumulated into a comprehensive knowledge of nature profoundly embedded in a sense of belonging. Such native knowledge necessitated acceptance of a limited horizon, which he delighted in. 'Let change be far from me,' he declared, 'that irresistible change must come is bitter indeed. Give me the old road, the same flowers – they were only stitchwort – the old succession of days.'

As I steadily and deliberately gathered my knowledge of the Cambridgeshire flora, I was mindful of the fact that this was not the countryside of my birth and childhood. I was a fleeting inhabitant of this landscape and I knew that I had no

intentions to remain here. There were other places I wished to visit, to which I might prefer to belong. My explorations of extinction had made me question how the origins of a person's understanding of nature affected their motives for safeguarding it. Even more so, I had come to recognize that different aspects of our efforts to survive were sometimes irreconcilable.

On the one hand, knowledge of nature might be inherently localized and limited in its scope, like that once possessed by the people of Whitby and still remembered by the Inuit elders. Those endowed with such intimacy with a particular place sensed that the natural world ordinarily altered through slow, almost imperceptible processes – 'the old succession of days'. One person's brief life afforded the luxury of natural knowledge gained at birth and confirmed through growing up in a single landscape. Successive generations could capture the incremental alterations occurring to such a place and the occasional hiccups of its seasonal variation, establishing the illusion of changelessness without shattering it. Such expertise would pass down through the generations, each new generation ensuring that the inherited knowledge still matched the reality of the world.

These traditions were destroyed when the environment changed too quickly, when societies grew too large or technological developments caused sudden deficiencies. Then the old knowledge was thrown into chaos, and a sense of approaching ruin only encouraged people to pull away from place and tradition, in the hope of stealing a march on extinction.

With the advent of the first industrialized societies, debates about the animal nature of humanity gave rise to the notion that one's birthplace was associated with primitiveness, whereas escape into both adulthood and elsewhere was the trajectory of civilization. Industrialization tended to dissolve rural traditions, along with dialects, vernaculars and nature lore of people largely enclosed in their own worlds. In industrialized societies, another kind of natural knowledge gained currency. Francis Bacon and similar thinkers considered the purpose of natural knowledge to be the governance of nature. This 'New Philosophy' of the seventeenth century in Britain grew into the Royal Society for the Improvement of Natural Knowledge, the scientific institute now known simply as the Royal Society. In 1890 the biologist Thomas Henry Huxley defined natural knowledge as a scientific understanding of the world. 'Our business was . . . to discourse and consider of philosophical enquiries, and such as related thereunto: − as Physick, Anatomy, Geometry, Astronomy, Navigation, Staticks, Magneticks, Chymicks, Mechanicks, and Natural Experiments.' These enquiries galvanized technological improvements that required new knowledge of the world.

Enterprises like the *Oxford English Dictionary* or *Webster's Thesaurus* documented the invention of new words in the English language. According to the *OED*, the word 'robot' was introduced in 1922, derived from the Czech word *robota*, for 'forced labour'. Later, as the technology of robotics advanced, it came to mean 'a machine capable of automatically carrying out a complex series of movements'. Meanwhile,

novel words predicted as future additions to the dictionary betrayed the emphasis of new knowledge: 'insourcing', 'made-for-mobile', 'pay by touch', 'softphone', as did those words already added, such as 'prime time', 'radiophysics', 'clonable', 'wire speed' and 'blogosphere'. These language changes signal the shift of natural knowledge in industrialized societies from the organic to the man-made environment.

This move from distinctive cultural knowledge born of the varied attributes of landscapes to the universal cultural knowledge of technologies available worldwide is akin to the disappearance of diversity in nature. The gardens of people's minds have been writ large in the physical realities of their surroundings. Landscapes bereft of their distinctiveness reflect human habits that have become indistinguishable, regardless of where people live.

The first page of my notebook was for the month of April at Fidwell Fen. My list of identifications read, 'Cuckoo flower, ground ivy, red dead nettle, white dead nettle, common field speedwell, common chickweed, lesser trefoil.' Later, in May, on my walks around Clayhithe, I added, 'Creeping buttercup, medium-flowered wintercress, silverweed, slender speedwell, nipplewort, honesty, herb Robert and greater periwinkle.' The flowers jotted down in my book were all common varieties, most especially the nettles, which have colonized great swathes of the British countryside. My discoveries reflected the spread of similar lifestyles across the world, undermining the idiosyncrasies of nature, increasing homogeneity. At the time of writing, over half of the nearly seven thousand unique

human languages are endangered, their traditional cultures carved by experience of special places now under threat.

Aristotle raised the spectre of equality both in his *Nicomachean Ethics* and in *Politics*. *Isos*, the Greek word for equality, was closer in meaning to the concept of fairness. It was the giving of what was due in correspondence with an individual's occupation or social situation. Equality, in this sense, was always relative. But Aristotle's notion of political justice suggested a different kind of equality, one born of participation in a common life. Justice could exist only among those who were free and equal in their capacities and intentions as citizens. This idea of political justice lent modern civilization its grounds for equal human rights. Following the horrors of the Second World War, the newly formed United Nations drafted and adopted the Universal Declaration of Human Rights in the winter of 1948. The thirty rights to which every person in the world was deemed to be entitled included:

Article 3. Everyone has the right to life, liberty and security of person . . .

Article 9. No one shall be subjected to arbitrary arrest, detention or exile . . .

And so on. Many of the articles were sustainable as universals across societies and time, but some articles raised a dilemma. Could all people have equal rights to some pursuits, if their destructiveness varied around the world? Could societies justify actions now known to cause short-term and long-term damage to the natural world, if the relative

damage depended on population size or the kind of technology? Article 27 stated that everyone had the right 'to share in scientific advancement and its benefits'. Could all humans have equal rights to a technology if this right, when exercised by every person, would devastate nature? The notion of equal and universal rights to all technologies enshrined in Article 27 depends on a kind of consistency across the Earth. It assumes that such rights can exist without relative impact in different landscapes and societies.

The diversity of landscapes, materials and life forms across the Earth has yielded an extraordinary range of products to aid human survival, but their universal application by all societies around the world, regardless of population size and their unique surroundings, requires an essential denial of the fragility of such diversity. Increasing the similarity of environments, as industrialized societies have done through the years, has eroded the natural variety of the world, largely through humanity's reliance on technological progress, now deemed a universal right of our species.

Occasionally, I speculated on whether scientists might find genetic ways to persuade societies and individuals to preserve nature. Knowledge of the human genome along with technologies like pre-implantation genetic screening could allow an interference at the biological level with human nature, an intrinsic transformation of our species that previously only natural selection could achieve. In designing our own nature, what would we choose to keep or abandon from our animal past? Could we eradicate the natural traits that

lead to the destructiveness of large, technological civilizations? Could we breed into children an altruistic sensitivity
to nature that might guard against its future ruin? Or the
self-command necessary to exist more contentedly with
finite resources? Leaving aside the moral conundrums of such
an intrusion, would genetic engineering produce different
results to the arbitrary trials of life? Would interfering with
children's genetic make-up be different to endowing them
with information from birth? Meeka said that the Inuit had
no hunting 'laws', for they believed that experience of nature
would moderate behaviour naturally. But as modern lifestyles
separate the majority of people from the natural world, such
mechanisms might never come into play.

I began my thinking for this book with an intense sympathy for nature but little or no understanding of it and a
growing sense of alarm at my own ignorance. While the tex-
tures and shades of my native English landscape were strongly
familiar to me, I had not grown up knowing its distinctive
plants and animals. As I concerned myself with the grim history of extinction, my findings began to alter my own behaviour – at first almost imperceptibly but later in conscious and
deliberate ways. I began to grasp my own place in the natural
landscape, particularly the landscape in which I was living.

In writing of his own interest in wild flowers, Richard Jefferies said that he did not recognize 'whence or why it was
joy'. In the nineteenth century, in Jefferies' lifetime, it was still
possible to revel blindly in this natural compulsion to enjoy
wildlife. But Jefferies' benevolent carelessness of experience

could not be mine. While my pleasure proceeded from an intrinsic desire to discover and understand my landscape, I knew why it was joy, and knew – more significantly – why it *had* to be joy. Although the beauty of wild flowers may have made me notice them, it was my sense of their imperilment that inspired my knowledge of them, and the more I knew of both the flowers and the wildlife around them, the more I was motivated in my occupation. This was a profoundly nostalgic way of behaving. It was recognition of the environmental destruction of the age into which I was born that provoked a further recognition of my own loss of any inherited understanding of the Earth from former generations. My observation of wild flowers and the pleasure I took in them was intimately bound to a conscious and moral reclamation of knowledge.

Growing up in Britain, the first nation in the world to become industrialized, I lived amid the ghosts of former wildness. While disappearances had occurred over the millennia as people became more and more skilled at exploiting and modifying nature, industrial processes accelerated their destructiveness by orders of magnitude. In the late 1980s, while I was still at junior school, the United Nations' Ad Hoc Group of Experts began discussing a possible international protocol to slow the rate at which humans detrimentally affected the natural world. By the time I returned from the Arctic, nature continued to be blighted, despite the efforts of conservation organizations and the promises of governments. The extinction of species remained many times higher than natural levels,

and diversity was still declining along with habitat loss. In the past century, Cambridgeshire alone had lost nearly seventy species of wild flower that formerly flourished in the county.

The ghost orchid was emblematic of all endangered wild flowers. This fragrant, leafless orchid, its flower a crookedness of pale pink and yellow, once grew in the beech woods of several English counties but was declared extinct in 2005. Last recorded in Britain in some woods in Buckinghamshire, it was discovered again several years later by a wild-flower enthusiast one autumn as he walked among some old oaks. With just one secret flowering each season in Britain, the orchid was more haunting than if it had been lost to extinction. It called to mind a few beautiful lines from Yeats's poem 'All Souls' Night':

> *A ghost may come*
> *For it is a ghost's right . . .*
> *His element is so fine,*
> *Being sharpened by his death.*

Our species' nutrition is given as the paramount reason to conserve and strengthen the variety of life on the Earth, as we require a rich source of food and our own capacity to yield this variety is limited. The other persuasive reasons given are our health, the security of our homelands against natural disasters, and our ability to exploit the materials of the Earth for manufacture. In this way, these arguments place the wildness of the Earth inside the garden of human concerns.

Underlying these justifications is the old presumption that living things and landscapes exist for our sake.

The laudable efforts of governments, conservationists and economists remain pointless without a real, reflective understanding of our behaviour as an animal, uncomfortable though that may be. For much of our history as a species, the pursuit of control has led us to dominate nature, but rationality and inventiveness have also allowed us to make sense of both the recent and the ancient past, steadily condensing information into a much more layered perspective on the natural world. A scientific understanding of nature has afforded insight into the limits of solely generational knowledge, not least because such knowledge relies on our presence. It has enabled us to appreciate our own eventual, unavoidable extinction. This humbling perspective encourages us to harness our rational skills and nostalgic sentiments to more benevolent, thoughtful ends. Our need to exploit nature for our own survival is balanced by an appreciation of the finiteness of our own species and the desire to study imperilled nature and try to salvage it.

The idea of the brutality of our animal nature, which is largely governed by intrinsic genetic traits, makes it hard to know to what extent we can alter human behaviour. However, regardless of the origins of our behaviour, we need to believe that we have the power to act differently. Firstly, by learning as children about our connection with nature and about the wild traits that continue to affect us, we might better anticipate which of these are likely to help or hinder

our progress. Such knowledge of human nature ought not to derive only from lessons in classrooms. The findings of science should be integrated with a personal natural knowledge gained by daily closeness to wildlife. This closeness originally derived from use but can now emerge from the importance and attraction of knowledge itself. Effective policies need to be built on these foundations.

A return to a more immediate alliance with nature is essential so that knowledge of how we threaten it can alter our behaviour – reawakening the sense that we live finite lives in a finite world. The spectre of overpopulation remains. For people to return to a reasonable knowledge of nature while still reliant on it for survival, our societies will have to agree on a means of fairly and peacefully controlling the size of the world population, along with the size of societies relative to the fragility of the environments in which they live. Some may dream of colonizing other planets. Such escapism ignores the fact that nature endures beyond our ideas or the Earth's bounds. We will simply perpetuate damage elsewhere. Others may look a long way forward to a time when our history as a species is fossilized in the Earth's annals like the remains of a scrapped car, a perspective that may blunt their belief in contemporary action. Instead, by closing the gap between the natural landscape and ourselves, a direct understanding of nature can govern people's motives, rather than responding to a world made in the image of human desires. By returning to a daily closeness with the natural world, and learning about species other than ourselves, we might strengthen the

moral sense that derives from our own origins in nature. My hope is that, in this way, nostalgia and inventiveness can come together to counteract our species' destructive tendencies.

I took the boat through the overgrown lock and up the narrow waterway of Wicken Fen, taking note of golden irises and water forget-me-nots like queens and paupers on the riverbank. The flitting psychedelic flight of a kingfisher left a vivid vein of blue across my vision. Two marsh harriers in black shadow fulfilled their tryst of appetite, the male slung beneath his mate as if the air was his bed, offering up his catch. Perched on the roof of the boat, my legs dangling into the engine room, my hand vibrating on the tiller, I had time to make these observations. I'd come to Wicken Fen precisely to bring wild nature to the fore.

Progress in a narrowboat is slow, especially through cramped or shallow waters – a brisk walker could easily outstrip me. The mind slows, too, thoughts drifting through the landscape like the languorous wake of the boat. I scanned the fields, softly umbering in the setting sun, and allowed my ideas to reel backwards and forwards between past, present and future, occasionally snagging on something physical, like a fly-fisher's line spun in and out. In the kitchen below deck, Ewan was preparing our evening meal, occasionally bobbing up at the bow to study something through his binoculars. We felt like émigrés among the wild inhabitants of this place, stripped back to the bare essentials of what we might have in common.

Living on the boat, I became more and more alert to the countless alterations of life around me. The first thing I had noticed was the hatching spiders, when it was not quite spring. Sunlight daggered the banks of the river but the light was pale as buttermilk, not strong enough to burn or to rouse. Traditionally, spring arrives in Britain as the sun crosses the celestial equator on 21 March. This is the vernal equinox, from the Latin *vernalis*, meaning 'spring', associated with *verno*, the Latin verb 'to bloom'. For most of my life, this date – automatically printed in calendars – was all that spring meant to me. But the river and its wildlife were challenging this view. The obstinate complexity of the natural life that surrounded me had ruptured the idea of spring into dozens of smaller, subtler seasons that I had never before perceived. For the first time in my life, I was living close enough to the elements to witness on a daily basis the arrhythmic strains of the natural world. A number of weeks before the traditional onset of spring, I experienced the thrill of discovering changes to which others around me seemed oblivious. People walking along the riverbank still bemoaned the winter and longed for its end. I wanted to open my windows and call out to them, 'Can't you see? It's already over! Spring is here!' Once the river had offered up this first secret, I was bound to the landscape by fierce curiosity. The presence of the spiders signalled the return of insects to the skyline of the greenery. I couldn't put a name to any of this diminutive, darting life but I was wide awake to its agitation among the green dimness of the hedges and banks. Questions popped into my mind.

From what kind of incubation had the various flying creatures emerged? And what kind of courtship could they hope to take to the air? I wanted to understand something of what I was seeing. The feeling was one of both elation and daring, each morning sprang up with possibilities.

In early English, the strong verb 'to spring' was *springan*, from which derived *sprengan*, the causative verb 'to break'. But the word 'spring' encompassed more than this: a sudden leap, the motive of an action, the emergence of life, the first stage of life, and the means of escape. All this meaning welling up through the ages from that first, pristine, violent word for a break. On this first spring on the river, I took to heart a break of sorts. As the wind reconsidered the world outside the cosy narrows of my boat and the distant sirens reminded me that I was moored between an urban and rural world, my mind was jolted out of passivity.

That afternoon, Wicken Fen told me two different stories: there was the landscape I could see and the landscape that had disappeared. Properly known as Wicken Sedge Fen, it is a flood plain bounded by clay banks and the watercourse through which I had cruised. When the Celtic tribes began populating Britain, the Cambridgeshire Fens were wild, salty marshes. The landscape remained this way until it was drained by Dutch engineers in the seventeenth century. Thomas Babington Macaulay referred to the inhabitants of the fens as 'a half-savage population . . . who led an amphibious life, sometimes wading, sometimes rowing, from one islet of firm ground to another'. Those living on the marshes could live

a subsistence lifestyle, supporting their family from a single boat or drained swamp, selling cheese, butter, fish and timber to London merchants. It was a hard life. In *A Tour through the Whole Island of Great Britain*, Daniel Defoe joked that some of these farmers married ten or fifteen women in their lifetime, as the wives often died early. Once a lush forest, the ancient woodlands were drowned as the last Ice Age retreated, forming a rich peat soil that was later harvested for fuel. Sedge colonized the land after the removal of the peat, which people then cut to thatch houses.

By the end of the nineteenth century, the disappearance of thatch in favour of tiles brought an end to sedge cutting. Wicken Fen became worthless, the preserve of naturalists. It was just an island of unwanted ground amid the wider fenlands. Drainage had made them highly profitable, the prime agricultural region of the country, producing a third of England's potatoes and a huge acreage of cereals. This transformation of the landscape caused considerable losses of species.

By the end of the nineteenth century, the bittern, spoonbill, greylag goose, marsh harrier, ruff, black-tailed godwit, avocet, black tern and Savi's warbler were lost from the landscape as breeding birds. The exquisite swallowtail butterfly became extinct due to the decline of its food source, the Cambridge milk parsley. The fen orchid disappeared with the cessation of peat cutting, and the delicate fen violet began to flower less often.

The moorings at Wicken Fen were before a small bridge at the end of the lode, beset by cow parsley. Nobody else was there. The blurred light of the evening was like a dark purply bruise. As night fell, man-made sounds dimmed, and a different communication became perceptible, as exotic as hearing a new language for the first time. The few evening visitors retreated from the Fen and the bird calls, the teasing sounds of insects, the whistle of tiny mammals replaced the footfall of humans. From the darkness came the muscular slap of fish on the water, the chirrups of the moorhens, the crackle of swans inside their beds of reeds, and the quavering nightly cryptogram of the barn owl. Thrillingly, for the first time in my life, I heard the boom of a bittern, its deep-throated, aspirated *who-who-who* echoing across the fen. A poem from the First World War, Francis Ledwidge's lament for the Irish nationalist Thomas McDonagh, who was executed a year before Ledwidge died in Flanders, told of the drowned sounds of nature amid his own extinction:

He shall not hear the bittern cry
In the wild sky where he is lain
Nor voices of the sweeter birds
Above the wailing of the rain.

Listening to the bittern's cry, Ewan told me that the strange hoarse song belonged to the male as he courted his mate. He happened to pass me his binoculars at just the moment that the bird took flight across the pond. For four or five stolen seconds, I watched the ordinarily shy bird with its mud-coloured coat of stars glide from sight. Although the wonderful and peculiar lowing of the bird could travel for miles, it seemed to have come from somewhere nearby. It was exciting to imagine us as neighbours. The bittern became increasingly rare across Britain as the reed beds where it fed and mated began to disappear through the intrusions of agriculture and the sea. Due to hunting and the drainage of the fens, it became extinct as a breeding bird in Britain in 1886. The keen eyes of an amateur birder called Emma Turner, who lived on a houseboat on Hickling Broad in Norfolk, spotted the first returning bird from the continent in 1911. Since then, the bittern has maintained a perilously tenuous existence in England, ill-fated by its reclusion among the reeds.

As I lay in the dark with the sound of ducks pecking the reeds on the underside of the boat, I had one thing on my mind. I wanted to see if I could find the fen violet, a wild flower endangered in England whose hint of pale purple as thin as moth wings made it the most beautiful plant I could

imagine. I wondered, too, as I drifted into sleep, why it was knowledge of wild flowers that most fired my mind. Carolus Linnaeus, the Swedish botanist who revolutionized the categorizing of life on Earth, argued that plants were fundamental to everything, with their regular cycle of blazing adolescence and fruitful death. They spring up, he said, they grow, they flourish, they ripen their fruit, they wither, returning to the earth again. Over the years, plants had come to stand in my mind for the spring of diversity, the flowering origins of the clashing, propagating mishmash of life. And so, at the end of all these journeys, it still bothered me that I had no favourite wild flower. Perhaps the fen violet would become the one closest to my heart, I reassured myself, if I could discover it.

Darwin visited Wicken many times, and most likely made an exciting find in the region. In a letter written in 1846 to Leonard Jenyns, he described his discovery of the crucifix ground beetle. 'I must tell you what happened to me on the banks of the Cam in my early entomological days,' he wrote. 'Under a piece of bark I found two carabi (I forget which) & caught one in each hand, when lo & behold I saw a sacred Panagæus crux major; I could not bear to give up either of my Carabi, & to lose Panagæus was out of the question, so that in despair I gently seized one of the carabi between my teeth, when to my unspeakable disgust & pain the little inconsiderate beast squirted his acid down my throat & I lost both Carabi & Panagæus!'

A daring blend of orange and black, the beetle is one of

Britain's endangered species, found at only three remaining haunts. Recently, someone spotted it at Wicken Fen, over fifty years since the last sighting. I felt heartened by the reappearance of the crucifix beetle. Perhaps the violet might return too, despite its absence the previous year. The following morning, I began my walk by enquiring in the Wicken Fen Visitor Centre if I might see the fen violet. Armed with their advice, I headed off into the nature reserve.

An obsolete and puzzling meaning of the word 'wilderness' is an area of ground in a large garden or park, planted with trees in fanciful forms such as a labyrinth. In his *Journal of a Naturalist,* John Knapp wrote in 1829 of a summer's day at Hampton Court, where 'on the opposite side of the palace there is a large space of ground called the Wilderness, planted and laid out by William III'. The idea of paradise originated in the Old Persian word *apairi-daeza*, an orchard enclosed by a wall. When translating the word 'garden' for the Septuagint, the scribes used *paradeisos*. Paradise was a fantasy of freedom from the strains of survival in nature. The Greco-Roman tradition celebrated a golden age in which humans lived in a blessed state, free from toil and deprivation. In Greek myth, after death heroes went to the Elysian Fields, where life was unencumbered by storm and plague. Descriptions of voyages blended myth and fantasy with the blurred edges of the known world. Sailors like Iambulus reputedly reached the wondrous Happy Isles. Here, he claimed, inhabitants lived in a state of blissful nature, in a landscape of permanent and effortless fecundity. In the *Odyssey*, Homer

described the gardens of Alcinous, on the island of the Phaea-
cians, whose trees fruited eternally, never showing the welts
of overripeness, and whose fields were gilt with corn that
grew without the labour of farmers. These were idealized
and imagined landscapes. These pagan myths of blessed and
abundant nature were seeded in the Christian vision of the
garden of Eden, where the first humans lived amid serene
and unthreatening nature. Almost immediately, people began
to debate the geographical location of this perfect place. In
the east, exclaimed Hippolytus in the third century, to be
rediscovered by the righteous and adventurous. At the cross-
roads of the Tigris and the Euphrates rivers, argued Bishop
Epiphanius. St Augustine, too, was convinced that Eden was
a historical place. How might people reach it again? When
Christopher Columbus sailed towards the Orinoco river on
his third voyage, he speculated that a vast summit or uncross-
able waters impeded his further progress. 'For I believe that
the earthly Paradise lies here,' he wrote, 'which no one can
enter except by God's leave.' Paradise became a garden into
which we could never hope to pass again.

From the Middle East, the tradition of the prince's garden
spread around the world as the Arabic rulers occupied new
lands. The famous Rusafa garden near Cordoba in Andalusia
was constructed as a nostalgic memory of the Emir's homelands in Syria. Such gardens were oases of life amid the relative poverty of a barren landscape and crippling heat. They
flowered in defiance of the deserts of the Middle East, where
nothing grew and drought always threatened. These gardens

were always hemmed-in worlds, governed by people, the first nature reserves, perhaps.

One of the earliest reserves, created in 1899, Wicken Fen was the place with the highest diversity of species in Britain. *The Entomologist*, edited by John Carrington in 1880, noted that 'Wicken still retains its virgin soil and flora, unspoilt by drainage or cultivation.' It became synonymous with hunts for rare butterflies and insects. 'An apparently good night frequently produces little or nothing, while sometimes those collectors who have had the patience or perseverance to stay through a wet and windy night find that suddenly the moths begin to come and many rarities are unexpectedly taken.' In 1895, an entomologist called Herbert Goss approached the newly instituted National Trust to save Wicken Fen and its exceptional range of species. At the close of the century, the Trust purchased several acres of the fen for ten pounds. It was a curious oasis in the black soil of Britain's principal area of farmland. The survival of endangered species like the crucifix beetle or rare wild flowers most probably depends on such sanctuaries for diverse life. But how can people gain native knowledge of them if they exist only within the confines of a nature reserve? The abundance of nature has been restricted to these sanctioned, bordered realms, into which people tip-toe within the allotted hours.

The light was beginning to go across the fen, gilding the tips of the rushes. I had failed to find the fen violet, but I didn't mind. I had made two pure and childlike discoveries of a more modest variety. Standing at a crossing of three paths, I

had stopped to orient myself. I could see the green brows of a stand of tall trees further along one path and, beyond these, a flickering stretch of water. Somewhere near here, I guessed, was the hiding place of the bittern I'd heard the night before. Taking this little alleyway down past some willows, I spotted a buff-brown bee that I'd seen once before in Cornwall. Its three tiers of reddish fur made it easy to identify. The books I had to hand told me it was not a bumblebee, but a solitary bee, a tawny mining bee that made its home alone in the earth. I liked the idea of it softly humming, hunkered in a dusty burrow, its minute heart drumming within the darkness of its body.

In my excitement, I somehow believed I would chance across the fen violet, as if enthusiasm alone might rush me towards its discovery, just as young children, in sickening desire, truly believe they might one day see a ghost. A little further along, I came across a flower similar to the delicate violet, with a bluish heart and rounded petals. My heart jolted. But as I flicked quickly through my wild-flower book, I realized it was only a relative of the fen violet and the common dog-violet, the flower my grandmother loved most. It resembled both but was touched with a little golden yoke of the summer to come. This was heartsease. It had other names in English, love-in-idleness, kiss-me-at-the-garden-gate, and was the flower whose juice Oberon squeezed on the eyes of Titania in *A Midsummer Night's Dream*, an amorous tonic to soften the hardest heart and renew admiration for the world. Even Darwin found the flower intoxicating. He undertook

a study of its propensity to shimmer through different hues when planted in other soils, slowly and stubbornly returning to its traditional colours by the summer's end. My guide classified it as *Viola tricolor*, the wild pansy, common and uninspiring, perhaps. But at that moment, I didn't want to savour what was dying out. I wanted to revel in the familiar, in something that I might be sure of greeting again. Not a ghost but an omen I could wait in hope for each spring.

Notes

Beginnings

2 'The entire class of birds': Charles Darwin (1859), *On the Origin of Species by Means of Natural Selection, or the Preservation of Varieties; and on the Perpetuation of Varieties and Species by Natural Selection*, Chapter X.

2 'A long lizard-like tail' : ibid.

3 'One of the strangest animals': Darwin, Charles (1839), *Narrative of the Surveying Voyages of His Majesty's Ships Adventure and Beagle*, Volume III, *Journal and Remarks 1832–1836*, Chapter V; reissued as *The Voyage of the Beagle*.

5 'Encountering the giant mammals': ibid., Chapter VIII.

6 'She's heavy/In the air': Melanie Challenger (2006), *Galatea*.

7 'From so simple a beginning': Darwin, *Origin of Species*, Chapter XV.

8 'Every great change of climate': Charles Lyell (1830), *Principles of Geology, Being an Attempt to Explain the Former Changes of the Earth's Surface, by Reference to Causes Now in Operation*.

9 'Life on earth has been frequently interrupted':

Georges Cuvier (1812), *Recherches sur les ossemens fossiles de quadrupèdes, où l'on rétablit les caractères de plusieurs espèces d'animaux que les révolutions du globe paroissent avoir détruites*, trans. Robert Kerr, *Essay on the Theory of the Earth*, 1813.

10   'The studies of elephant bones': Cuvier delivered this statement in 1796 to the Institut National de France in Paris, published as *Mémoires sur les espèces d'elephants vivants et fossils*, 1800.

10   'If this animal once existed': Thomas Jefferson (1799), 'Megalonyx', *American Philosophical Society*, Volume IV.

11   'Whether a savage or a nobleman': John Fleming (1828), *A History of British Animals, Exhibiting the Descriptive Characters and Systematical Arrangement of the Genera and Species of Quadrupeds, Birds, Reptiles, Fishes, Mollusca, and Radiate of the United Kingdom, Including the Indigenous, Extirpated, and Extinct Kinds, Together with Periodical and Occasional Visitants*.

11   'Man is everywhere a disturbing agent': George Perkins Marsh (1886), *Man and Nature, or, Physical Geography as Modified by Human Action*, Introduction.

The First Peregrination

1   Wild flowers

20   'Ding-dong! Merry, merry, go the bells': Henry Kirk White (1785–1806), 'The Dance of the Consumptives', *The Poetical Works of Henry Kirk White*.

25   'An autobiography in which he described the flowers':

John Porter (1838), *John Porter of Kingsclere: An Autobiography*.

32 'The poorest in the county': J. M. Neeson (1993), *Common Right, Enclosure and Social Change in England*.

33 'Communal husbandry began to alter': see E. L. Jones (1965), 'Agriculture and Economic Growth in England, 1660–1750', *Journal of Economic History*.

37 'Lighten any check': Darwin, *Origin of Species*, Chapter III.

41 'Whether as convalescence or compensation': Edgar Thurston (1922), *A Supplement to F. Hamilton Davey's 'Flora of Cornwall'*. This contains a short biography of Frederick Davey by Chambre Vigur.

42 'The green vale': Dorothy Wordsworth, quoted in Geoffrey Grigson (1984), *The English Year from Diaries and Letters*.

44 'Roses red and white': Thomas Hood (1827), 'I Remember, I Remember', *The Plea of the Midsummer Fairies*.

44 'Stand on the highest pavement': T. S. Eliot (1917), 'La Figlia Che Piange', *Prufrock and Other Observations*.

45 'Remember and foreknow:': Edward Muir (1979), 'The Breaking', *Times Literary Supplement*.

46 'We have no festival, nor procession, nor ceremony': Henry David Thoreau (1854), *Walden; or, Life in the Woods*.

46 'Socrates and Euthydemus debated': see Xenophon

(c. 431–350 BCE), *The Memorabilia*, trans. Henry
Dakyns, 2007.

47　'Their most rich and fertile fruits': Marcus Tullius
Cicero (45 BCE), *De Natura Deorum*, trans. C. D. Yonge,
On the Nature of Gods, 1913.

48　'Since only we could rationalize': Thomas Aquinas
(1264), *Summa Contra Gentiles*, trans. Vernon J. Bourke,
1973, Chapter 78.

48　'Subduing or cultivating the earth': John Locke (1689),
Two Treatises of Government.

53　'The extinction of the passion': Thomas Malthus
(1798), *A Principle of Population; or A View of its Past and
Present Effects*.

53　'Wholly cultivated and improved': William Godwin
(1820), *Of Population: An Enquiry Concerning the Power of
Increase in the Numbers of Mankind*.

53　'But if a man endeavour to establish': Francis Bacon
(1620), *Novum Organum*, trans. James Spedding, 1858.

2　Tin

56　'Wrætlic is þes wealstān': Michael Alexander (2006),
The Earliest English Poems.

57　'Walls stand, wind-beaten': ibid.

58　'A continuous vibration of animal spirits': Johannes
Hofer (1688), *Medical Dissertation on Nostalgia*, trans.
Carolyn Kiser, 1934.

60　'*Leviathan*': Thomas Hobbes (1651), *Leviathan or the*

Matter, Forme and Power of a Common Wealth Ecclesiasticall and Civil.

63 'The inhabitants, he said, were busy preparing tin': F. Halliday (1959), *A History of Cornwall.*

64 'Behind such inventions': J. Farey (1971), *Treatise on the Steam Engine.*

64 'We avail ourselves': ibid.

65 'Ho! Brothers, ho!': anon., 'Song of the Granitic Grandee'.

73 'Orient mine, Trespuntas': letter from a Tucking mill miner to his family, *The Cornish in Latin America*, Exeter University.

3 Ghosts

78 'Mechanical memory': Maine de Biran (1804), 'Sur l'influence de l'habitude', trans. Stephen Kosslyn and Richard A. Anderson, *Frontiers of Cognitive Neuroscience*, 1995.

84 'Streets were often so narrow': Fortescue Hitchins and Samuel Drew (1824), *The History of Cornwall, from the Earliest Records and Traditions, to the Present Time.*

84 'Necessity rather than choice': ibid.

86 'Grief as a universal instinct': see John Bowlby (1969), *Attachment and Loss.*

90 'Theophrastus said of Cyrene': see J. Donald Hughes's excellent article, 'How the Ancients Viewed

Deforestation', *Journal of Field Archaeology*, 1983, Volume 10.

94   'The most skilled interpreter of dreams': Aristotle (350 BCE), *De Somno et Vigilia*, trans. J. I. Beare, 1912.

95   'Through the dewy dark on noiseless wings': Ovid (8 ACE), *Metamorphoses*, trans. A. D. Melville, 1986.

96   'A net so close meshed': Callum Roberts (2007), *The Unnatural History of the Sea: The Past and Future of Humanity and Fishing*.

The Second Peregrination
4   *Whales*

101 'I can see no difficulty': Darwin, *Origin of Species*, Chapter VI.

101 'Picture some primitive, marsh-haunting': William Henry Flower (1898), *Essays on Museums and Other Subjects Connected with Natural History*.

108 'The whales are creatures': Willem Van der Does (1934), *Storms, Ice and Whales: The Antarctic Adventures of a Dutch Artist on a Norwegian Whaler*, reissued 2003.

110 'All a long dream': D. Davies, *Diary*, British Antarctic Survey archives, reference AD6/15/40.

115 'It is a kind of sad farewell': J. B. Priestley (1947), *The Linden Tree*.

117 'Rid itself of the ill-omened debris': F. T. Marinetti (1909), *The Futurist Manifesto*, trans. James Joll, 1961.

119 'During the First World War': Fritz Thyssen (1941), *I Paid Hitler*, trans. Emery Reves.

122 'Hitler liked to say': Albert Speer (1969), *Erinnerungen*, trans. Richard and Clara Winston, *Inside the Third Reich*, 1970.

124 'Then one man spoke in his dream': quoted by W. G. Sebald (2003), *On the Natural History of Destruction*, trans. Anthea Bell.

129 'Colin Clark argued': C. W. Clark and R. Lamberson (1982), 'An Economic History and Analysis of Pelagic Whaling', *Marine Policy*.

5   Ice

135 'Sleeping in my bed, strange thoughts': Lal Waterson (1972), 'Fine Horseman', recorded on the *Bright Phoebus* album. Waterson lived and worked near Whitby, North Yorkshire.

136 'Exploring around our country's shipping ports': see www.chrisjordan.com.

139 'Ah wretch! said they': S. T. Coleridge (1798), *Lyrical Ballads*.

147 'Europe can afford romanticism': Paul Collier, *Guardian*, 22 August 2000.

150 'This nation despises agriculture': Gerald of Wales, 'Giraldi Cambrensis Opera', ed. J. F. Dimock (1891).

150 'Everything that concerns my needs': Jean Jacques Rousseau (1781), *Reveries of the Solitary Walker*, trans. Peter France, 1979.

152 'For the brute': Jean Jacques Rousseau (1755), *A Discourse upon the Origin and Foundation of the Inequality among Mankind*, trans. G. D. H. Cole, 1920.

156 'In a letter': Frederick Jones (1944), *The Letters of Mary W. Shelley*.

157 'Thus in the raging': Gavin Kelly (2004), 'Ammianus and the Great Tsunami', *Journal of Roman Studies*, 94.

161 'A lumbering lubbard loitering slow': Hermann Melville (1888), 'The Berg: A Dream', *John Marr and Other Sailors*.

170 'A new explosive': Harry Truman (1955), *Memoirs*. President Truman claimed that Secretary Henry Stimson made the comment on 12 April 1945, soon after he was sworn in.

170 'The raindrops were big and black': British Library Sound Archive Catalogue, *Hiroshima Witnesses Videos*, Hiroshima Peace Culture Foundation, reference C1050.

176 'Improve our agriculture': Tench Coxe (1787), 'Prospects for American Manufacturing', *View of the USA*.

6　Savages

184 'They might have continued to live': Robert Young (1905), *From Cape Horn to Panama*.

185 'During the progress of their education': John W. Marsh and W. H. Stirling (1883), *The Story of Commander Allen Gardiner, R.N., with Sketches of Missionary Work in South America*.

189 'It was without exception' : Darwin, *Voyage of the Beagle*, Chapter X.

190 'It was both laughable and pitiable': ibid.

190 'He was ashamed of himself': ibid.

194 'The single inconvenience': quoted by M. W. Holdgate, 'Man and Environment in the South Chilean Islands', *Geographical Journal*, 1961.

196 'Eventually rising from earthly': Saint Augustine of Hippo (426), *De Civitate Dei*, trans. Marcus Dods, *City of God*, Book X, Chapter 14.

197 'Through the advance of languages': Anne-Robert-Jacques Turgot (1750), *Notes of Universal History*, trans. William Walker Stephens, 1895. See Robert Nisbet's excellent *History of the Idea of Progress*, 1980.

199 'The most salient fact': J. Huxley (1942), *Evolution: The Modern Synthesis*.

200 'The more I have seen of these and other primitive tribes': Charles Wellington Furlong (1917), 'Some Effects of Environment on the Fuegian Tribes', *Geographical Review*.

201 'Stout ankles and squat, short-toed feet': H. T. Hammel (1960), 'Thermal and metabolic responses of the Alacaluf Indians to moderate cold exposure', technical report by Wright Air Development Center of the US Air Force.

202 'Whole bodies of men': Ferguson, *History of Civil Society*.

203 'Rudimentary parts': Darwin, *Origin of Species*, Chapter III.

204 'Injurious characters': Charles Darwin (1878), *Descent of Man*, Chapter V.

205 'Man can resist with impunity': Darwin, *Origin of Species*, Chapter VII.

206 'Dogs introduced to South America': Lyell, *Principles of Geology*, Chapter II.

211 'I am sometimes afraid': Francis Darwin, ed. (1887), *The Life and Letters of Charles Darwin*, Volume 1.

211 'I often think of the Garden at home as a paradise': ibid.

The Third Peregrination

7   *Bones*

221 'It remained an unpredictable endeavour': Charles Bowden (2004), *The Last Fisherman*, ITV.

222 'Nowadays there isn't many cobles left': Sarah Macmillan (2008), *The Fishing Town of Redcar*.

226 'When the blubber was boiled in the Arctic': William Scoresby (1820), *An Account of the Arctic Regions with a History and Description of the Northern Whale-Fishery*.

226 'In consequence of this event': ibid.

234 'The degree in which any man possesses': Francis Jeffrey (1826), 'Review of George Combe's *A System of Phrenology*', *Edinburgh Review*, 44.

238 'It remains true that nearly every respectable attribute': John Stuart Mill (1874), 'On Nature', *Three Essays on Religion: Nature, the Utility of Religion and Theism*.

238 'Either it is right that we should kill': ibid.

240 'Wake up, Cesare': Robert Wiene (1920), *Das Kabinett des Doktor Caligari*, screenplay by Hans Janowitz and Carl Mayer.

241 'I was haunted by the picture': Sigmund Freud (1915), *The Interpretation of Dreams*, trans. A. A. Brill.

243 'The society of the spectacle': Guy Debord (1967), *La Société du Spectacle*, trans. Fredy Perlman, 1970.

8 Tundra

246 'Will the Inuit disappear': see Milton M. R. Freeman (2000), *Endangered Peoples of the Arctic*.

249 'We have room for but one flag': Theodore Roosevelt speaking on immigration in 1919.

250 'Few future events are more certain': Lyell, *Principles of Geology*.

251 'It hath ever been the use': Edmund Spenser (1596), *A View of the Present State of Ireland*.

255 'The worlds in which societies live': Edward Sapir (1929), 'The Status of Linguistics as a Science', *Language*, 5.

257 'A thing composite and strange': W. D. Howells (1896), *Yekl: A Tale of a New York Ghetto*.

268 'Fragments of boats': Ari þorgilsson, *Islendingabok*, composed in the twelfth century.

270 'It would be hazardous and arbitrary indeed': Immanuel Kant (1764), *Observations on the Feeling of the Beautiful and Sublime*, trans. John T. Goldthwait, 1961.

275 'The first year I went to Cumberland gulf': see *From*

the Deep of the Sea; Being the Diary of the Late Charles Edward Smith, M.C.R.S., Surgeon of the Whale-ship Diana, of Hull.

275 'When the Eskimos who have spent the summer inland': Franz Boas (1888), 'The Central Eskimo', *Bureau of American Ethnology.*

Endings

290 'Without conscious thought of seasons': Richard Jefferies (1885), 'Wildflowers', *The Open Air.*

291 'The cave they lodged in': Adam Smith (1762), *Of the Origin and Progress of Language.*

293 'To discourse and consider of philosophical enquiries': Thomas Henry Huxley (1890), *On the Advisableness of Improving Natural Knowledge.*

297 'Whence or why it was joy': Jefferies, *The Open Air.*

299 'A ghost may come': William Butler Yeats (1928), 'All Souls' Night', *The Tower.*

304 'A half-savage population': Thomas Babington Macaulay (1848), *History of England from the Accession of James II.*

307 'He shall not hear the bittern cry': Francis Ledwidge (1917), 'Lament for Thomas McDonagh', *Songs of Peace.*

308 'Under a piece of bark I found two carabi': Charles Darwin, letter to Leonard Jenyns, 17 October 1846.

Acknowledgements

I must thank first my kind and unflaggingly positive agent
Jessica Woollard, and the Marsh Agency. Many of the ideas in
this book owe a great debt to the research and ideas of the
former Arts and Humanities Research Council's Centre for
the Evolution of Cultural Diversity, University of London,
and I thank Dr James Steele and Manu Davies in particular for their support. I am similarly indebted to the British
Antarctic Survey, and especially John Shears, the crew of the
RRS *James Clark Ross* under Captain Graham Chapman, and
my comrades on the voyage, in particular Jerry Armour, Phil
Coates, Paul Dennis, Jim Ditchfield, Ewan Edwards, Callum
Hunter, Dr Ruth McCabe, Mandy McEvoy, Rob Murdoch,
Claire Waluda and David Wheeler, for making it so inspirit-
ing; thanks also to David Walton for his notes on an early draft
and to Joanna Rae for her assistance in the BAS archive. My
travel was generously funded by the Arts Council of England
and the British Council Darwin Awards.

A number of individuals offered advice, guidance and
places to stay, and indispensible perspectives both formal and
informal on their lives in the Arctic. I am sincerely grateful
to Bert Rose and Markus Wielcke, and to Leesee Papatsie,

Simionie Keenainak, Billy Etooangat and Meeka Mike, for their sense of humour, trust and memories; and especially to Juavie Alivaktuk, for taking me out on to the land in the depths of winter while I was heavily pregnant, keeping me and my unborn safe, and teaching me and my husband how to fish for charr in the ice. I believe a little bit of Arctic magic went into my child that day.

A number of thought-provoking discussions arose among members of the Hans Rausing Endangered Languages Programme at the School of Oriental and African Studies, University of London, especially with Professor Peter Austin, which were helpful in shaping some of the ideas in this book.

I thank two long-standing friends whose encouragement has made a difference over the years, Dr Sasha Norris and Eitan Buchalter, and a new kindred thinker, Caspar Henderson, for moral support during the writing-up; the fantastic Greco-Clayton clan, for wine, humour, and acting *in loco familiae* in New York; friends in Cambridge who helped me and made the wine taste sweeter, Dacia Viejo-Rose and Benjamin Morris, and everyone at Clare Hall College – the writing desk made an immeasurable difference in lieu of a room of my own; Doug and Jenny at Ding Dong Mine Cottages for the cabin, and my father-in-law, Roger, for finding it in the first place; Professor Paul Thompson for my inspirational stay on Eynhallow; Dr Rodrigo Hucke-Gaete and everyone on the 2008 field season in Melinka for my first sighting of a blue whale; and Gillian and all the Newmans, for innumerable acts of neighbourly kindness in Cromarty.

I am particularly appreciative of the intelligent and enthusiastic editing of Michal Shavit and Daphne Tagg, and for the assistance of everyone at Granta Books. I am deeply grateful for such a supportive team.

Thanks to my father, John, and sister, Tamsyn, for huge amounts of stamina and belief over the years. Finally, gratitude to my beautiful and spirited son, Gabriel Suilven Wakefield, for coming into the world in the midst of writing this, exhausting me and reminding me exquisitely of nature each day that I struggled to write. To my mother, Pip, for overwhelming acts of generosity and patience, without which I could not have finished the book. Lastly to my husband, Ewan, for a shared passion for the natural world and a passion for each other. I dedicate this book to the three of you, mother, husband, son, with inexpressible love.

Illustrations Acknowledgements

17   David Wheeler

18   Melanie Challenger

19   Melanie Challenger

20   Melanie Challenger

21   Imperial War Museums Collection, MH29427

22   David Wheeler

23   Melanie Challenger

24   Melanie Challenger

25   Doran Brothers, with permission of the Whitby
     Literary and Philosophical Society

26   Doran Brothers, with permission of the Whitby
     Literary and Philosophical Society

27   William Scoresby, 1820

28   Frank Meadow Sutcliffe, with permission of the
     Sutcliffe Gallery, Whitby

29   Melanie Challenger

30   Melanie Challenger

31   A. G. McKinnon, Canada. Department of Indian
     and Northern Affairs/Library and Archives Canada/
     PA-101939

32   G. E. Briggs, Godwin Collection, Clare College,
     Cambridge